Introduction to the
THEORY OF THIN SHELLS

Introduction to the THEORY OF THIN SHELLS

H. MØLLMANN

Department of Structural Engineering
Technical University of Denmark

A Wiley–Interscience Publication

JOHN WILEY & SONS
Chichester · New York · Brisbane · Toronto · Singapore

British Library Cataloguing in Publication Data:

Møllmann, H.
Introduction to the theory of thin shells.
1. Shells (Engineering)
I. Title
624.1'7762 TA660.S5

ISBN 0 471 28056 9

Typeset by Preface Ltd, Salisbury, Wilts.
and printed in Great Britain by
The Pitman Press, Bath, Avon

CONTENTS

PREFACE

Thin shell structures are used in many branches of technology, such as building construction, mechanical engineering, shipbuilding, chemical engineering, aerospace engineering, and nuclear reactor engineering. The theory of shells is therefore an important subject in structural mechanics.

There is a great deal of literature on the theory of shells. However, most of the existing books on the subject are comprehensive treatises rather than textbooks, and some of the best known of them are rather old. There therefore seems to be a need for a fairly brief, modern introduction to the theory of shells.

It is the purpose of this book to provide such an introduction which can be used as a textbook for senior undergraduate or postgraduate students in structural or mechanical engineering. The book is based on a set of lecture notes prepared by the author for a one-semester course on shell theory given to fourth- or fifth-year undergraduate students in structural engineering at the Technical University of Denmark. (In this course, the final chapter on tensor methods is omitted.)

The first chapter is devoted to the general theory of shells. An introductory section deals with the results from the theory of surfaces (including the equations of Gauss and Codazzi) that are essential to the following derivation of shell theory. The general, linear theory of thin, elastic shells is then developed. Non-tensorial notation is used (lines of curvature coordinate curves), since tensor calculus is not included in the normal syllabus for engineering students. The shell is regarded as a two-dimensional body, and the fundamental importance of the principle of virtual work is emphasized. The resulting governing equations are those of the Koiter–Sanders theory.

The second chapter deals with axisymmetric bending of shells of revolution. The general, governing equations are derived and applied to spherical, conical, and cylindrical shells. Analytical solutions are derived by means of matrix methods which are convenient for numerical calculations. Geckeler's approximate solution for shells of revolution is also treated.

The theory of shallow shells is treated in the third chapter. The governing equations of shallow shells are derived by making appropriate simplifications in the equations of the general theory. The analytical solution for a shallow shell over a rectangular plan is derived, and the connection with Donnell's theory for circular cylindrical shells is discussed. The analytical solution for axisymmetric bending of shallow spherical shells is also derived.

The fourth and final chapter presents a derivation of general, linear shell theory in tensor notation. A student who has read the first three chapters of the book may, at a later stage, wish to familiarize himself with the modern tensor treatment of shell theory (perhaps in connection with postgraduate work). This last chapter is intended to facilitate this task for the student since it employs the kinds of methods and notation with which he is already familiar from the previous chapters.

Each of the first three chapters is followed by a set of problems, which may be used by the reader to check his understanding of the subject matter treated in that chapter. In a few cases, the problems bring out additional details that are not treated in the main body of the text.

Most of the material presented in the book is well known, but there may be a few original contributions in connection with the treatment of boundary conditions in the general theory, and the compact matrix formulation of the analytical solutions.

It is assumed that the reader has a good grounding in the theory of structures and that he is familiar with the elements of the classical theory of elasticity. The mathematical techniques employed in the first three chapters are reasonably elementary, and only a knowledge of calculus (including differential equations) and vector and matrix algebra is required. However, in the fourth chapter, it is also assumed that the reader is familiar with the elements of tensor calculus.

Although the book is primarily written as a textbook, I believe that it may also be of interest to practising stress analysts who wish to obtain some general background knowledge of shell theory to enable them to delve further into the literature on shells. For instance, a person who has studied the first three chapters of the present book will be able to read and understand the literature on shells in non-tensorial notation (such as the standard works listed in the first section of the bibliography), while the reader of the whole book will also be able to follow the literature on linear shell theory in tensor notation.

It should be noted that certain areas within the field of shell theory are not treated. These include finite displacements and the theory of stability of shells, inelastic (e.g. plastic) behaviour of shells, vibrations of shells, and numerical methods for the solution of shell problems (such as the finite element method and finite difference methods). The inclusion of some or all of these subjects would have increased the size of the book far beyond the bounds of a fairly brief introduction.

I am indebted to several persons for their help in the writing of this book. Dr M. W. Bræstrup, Dr J. Kirk and Mr Carl Pedersen, M.Sc., read Chapters 1 and 4 of the manuscript and made several valuable suggestions for improvements, and Mrs Pauline Katborg corrected and improved the first version of the translation.

I also wish to express my sincere gratitude to my wife for typing most of the translation and for her patience and understanding during the period of my work on this book.

Finally, I wish to thank the publisher, John Wiley & Sons Ltd, and Mr James Cameron, Publishing Editor, for their contributions to the successful completion of this book.

Lyngby, December 1980 H. MØLLMANN

1

GENERAL THEORY OF SHELLS

1.1 Introduction

A thin shell is a body that is bounded primarily by two closely spaced curved surfaces. More precisely, the geometry of a shell can be described in the following manner. We consider a surface in three-dimensional space determined by the parametric representation

$$\mathbf{r} = \mathbf{r}(\theta_1, \theta_2), \tag{1.1}$$

where $\mathbf{r}$ is the position vector from the origin O to points on the surface, and the domain of definition of the parameters is a closed region Ω in the (θ_1, θ_2)-plane. We assume that there is a one-to-one correspondence between the pairs of numbers (θ_1, θ_2) belonging to Ω and the points of the surface. Let $\mathbf{a}_3(\theta_1, \theta_2)$ denote the unit normal vector to the surface, corresponding to the values (θ_1, θ_2) of the parameters, and let

$$h = h(\theta_1, \theta_2), \qquad \text{where} \qquad h > 0, \tag{1.2}$$

be a given function of (θ_1, θ_2). We now consider the closed region S in space, the points of which are given by the position vectors

$$\mathbf{r}(\theta_1, \theta_2) + z\mathbf{a}_3(\theta_1, \theta_2) \tag{1.3}$$

where $(\theta_1, \theta_2) \in \Omega$ and $|z| \leq \frac{1}{2}h(\theta_1, \theta_2)$.

Consider a three-dimensional body whose particles occupy the region S at a certain instant. This body is called a *shell*, the surface (1.1) is called the *middle surface*, and h is called the *thickness* of the shell (see Fig. 1).

We assume that $\frac{1}{2}h < |R_{\min}| \neq 0$, where $R_{\min}$ is the numerically smallest radius of curvature of the middle surface (this condition ensures a one-to-one correspondence between (θ_1, θ_2, z) and the points of S).

A shell is called *thin* when the thickness is small compared with the other dimensions of the shell.

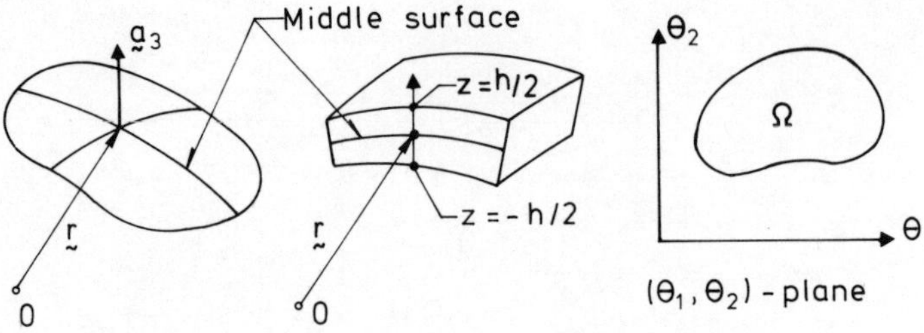

Fig. 1. Geometry of shell

1.2 Some results from the theory of surfaces

In this section we derive some important results from the theory of surfaces which will be used frequently in the following.

If we put $\theta_2 = c_2 =$ constant in the equation of the middle surface (1.1) we obtain the vector function $\mathbf{r} = \mathbf{r}(\theta_1, c_2)$ of the single independent variable θ_1, and this is the equation of a curve on the surface.

If we successively give c_2 a series of constant values, we obtain a family of curves which are called the θ_1-curves. In the same way we have another family of curves, the θ_2-curves, given by $\theta_1 = c_1$. In view of the assumed one-to-one correspondence between the values of the parameters (θ_1, θ_2) and the points of the surface, only one curve of each of the two families of curves will pass through each point of the middle surface. It is well known that (θ_1, θ_2) are called *curvilinear coordinates* of the surface, and the θ_1- and θ_2-curves are called the *coordinate curves*.

The *base vectors* $\mathbf{a}_1$ and $\mathbf{a}_2$ of the surface are defined by

$$\mathbf{a}_1 = \partial\mathbf{r}/\partial\theta_1, \qquad \mathbf{a}_2 = \partial\mathbf{r}/\partial\theta_2,$$

or briefly

$$\mathbf{a}_\alpha = \partial\mathbf{r}/\partial\theta_\alpha, \tag{1.4}$$

where we have introduced the convention that *greek subscripts* represent the numbers 1, 2. It will be seen that the base vectors $\mathbf{a}_1(\theta_1, \theta_2)$ and $\mathbf{a}_2(\theta_1, \theta_2)$ are parallel to the tangents to the θ_1- and θ_2-curves, respectively, through the point of the middle surface with coordinates (θ_1, θ_2) (see Fig. 2). It follows from (1.4) that we have

$$\mathbf{a}_{\alpha,\beta} = \frac{\partial^2\mathbf{r}}{\partial\theta_\alpha\partial\theta_\beta} = \mathbf{a}_{\beta,\alpha}, \tag{1.5}$$

where we have introduced the *comma notation* to denote partial derivatives with respect to θ_α, i.e.

$$(\ )_{,\alpha} = \partial(\ )/\partial\theta_\alpha. \tag{1.6}$$

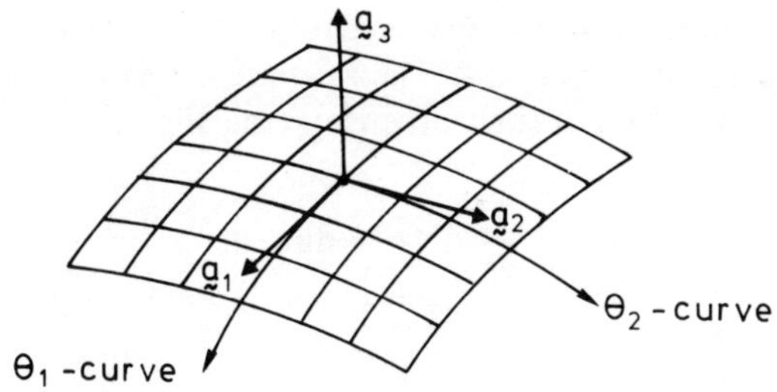

Fig. 2. Base vectors and coordinate curves

The infinitesimal vector connecting two points on the surface with coordinates θ_α and $(\theta_\alpha + \mathrm{d}\theta_\alpha)$ is given by

$$\mathbf{dr} = \frac{\partial \mathbf{r}}{\partial \theta_1}\,\mathrm{d}\theta_1 + \frac{\partial \mathbf{r}}{\partial \theta_2}\,\mathrm{d}\theta_2 = \sum_{\alpha=1}^{2} \mathbf{a}_\alpha\,\mathrm{d}\theta_\alpha. \tag{1.7}$$

The length ds of this vector is therefore determined by

$$(\mathrm{d}s)^2 = \mathbf{dr} \cdot \mathbf{dr} = \sum_{\alpha,\beta} \mathbf{a}_\alpha \cdot \mathbf{a}_\beta\,\mathrm{d}\theta_\alpha\,\mathrm{d}\theta_\beta. \tag{1.8}$$

Introducing the notation

$$a_{\alpha\beta} = \mathbf{a}_\alpha \cdot \mathbf{a}_\beta = a_{\beta\alpha}, \tag{1.9}$$

for the scalar products of the base vectors, (1.8) can be written in the form

$$(\mathrm{d}s)^2 = \sum_{\alpha,\beta} a_{\alpha\beta}\,\mathrm{d}\theta_\alpha\,\mathrm{d}\theta_\beta. \tag{1.10}$$

The quadratic form on the right-hand side of the equation, which determines the line element ds on the surface, is called the *first fundamental form* of the surface.

It was mentioned previously that the unit normal vector to the surface at a point with coordinates (θ_1, θ_2) is called $\mathbf{a}_3(\theta_1, \theta_2)$. The direction of $\mathbf{a}_3$ is determined in such a way that the vectors $\mathbf{a}_1$, $\mathbf{a}_2$, $\mathbf{a}_3$, in that order, form a right-handed system. It follows that

$$\mathbf{a}_3 = \frac{\mathbf{a}_1 \times \mathbf{a}_2}{|\,\mathbf{a}_1 \times \mathbf{a}_2\,|}. \tag{1.11}$$

where $|\,\mathbf{a}_1 \times \mathbf{a}_2\,|$ denotes the length of the vector $\mathbf{a}_1 \times \mathbf{a}_2$. Since $\mathbf{a}_1$ and $\mathbf{a}_2$ are vectors parallel to the tangent plane, and $\mathbf{a}_3$ is a normal vector, we have at any point of the middle surface that

$$\mathbf{a}_3 \cdot \mathbf{a}_\beta = 0. \tag{1.12}$$

Differentiating this equation with respect to θ_α we get

$$\mathbf{a}_{3,\alpha} \cdot \mathbf{a}_\beta + \mathbf{a}_3 \cdot \mathbf{a}_{\beta,\alpha} = 0. \tag{1.13}$$

Consider a point on the middle surface with coordinates θ_α and a unit vector **t** in the tangent plane at this point. The *normal curvature* associated with the direction determined by **t** is then given by (see Fig. 3)

$$\frac{1}{R}\,\mathrm{d}s = -\mathrm{d}\mathbf{a}_3 \cdot \mathbf{t},$$

or

$$\frac{1}{R} = -\frac{\mathrm{d}\mathbf{a}_3}{\mathrm{d}s} \cdot \mathbf{t}. \tag{1.14}$$

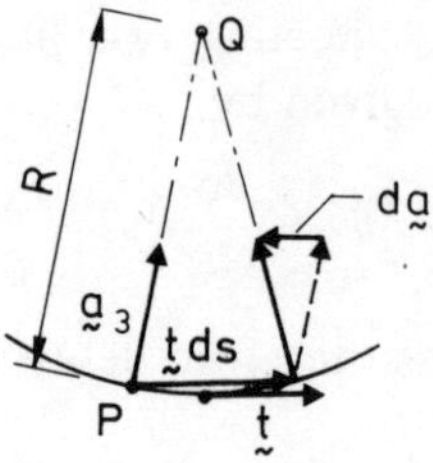

Fig. 3. Normal curvature

By using (1.7) we find that

$$\mathbf{t} = \frac{\mathrm{d}\mathbf{r}}{\mathrm{d}s} = \sum_\beta \mathbf{a}_\beta \frac{\mathrm{d}\theta_\beta}{\mathrm{d}s}. \tag{1.15}$$

We also have

$$\frac{\mathrm{d}\mathbf{a}_3}{\mathrm{d}s} = \sum_\alpha \mathbf{a}_{3,\alpha} \frac{\mathrm{d}\theta_\alpha}{\mathrm{d}s}. \tag{1.16}$$

Substituting these expressions into (1.14) and using (1.10) we derive

$$\frac{1}{R} = \frac{-\sum_{\alpha,\beta} \mathbf{a}_{3,\alpha} \cdot \mathbf{a}_\beta \,\mathrm{d}\theta_\alpha \,\mathrm{d}\theta_\beta}{\mathrm{d}s^2} = \frac{-\sum_{\alpha,\beta} \mathbf{a}_{3,\alpha} \cdot \mathbf{a}_\beta \,\mathrm{d}\theta_\alpha \,\mathrm{d}\theta_\beta}{\sum_{\alpha,\beta} a_{\alpha\beta} \,\mathrm{d}\theta_\alpha \,\mathrm{d}\theta_\beta}. \tag{1.17}$$

We now define the quantities $b_{\alpha\beta}$ in the following manner

$$b_{\alpha\beta} = -\mathbf{a}_{3,\alpha} \cdot \mathbf{a}_\beta = \mathbf{a}_3 \cdot \mathbf{a}_{\beta,\alpha} = \mathbf{a}_3 \cdot \mathbf{a}_{\alpha,\beta} = b_{\beta\alpha}, \tag{1.18}$$

in which (1.13) and (1.5) have been used. Equation (1.17) can now be written in the form

$$\frac{1}{R} = \frac{\sum_{\alpha,\beta} b_{\alpha\beta} \,\mathrm{d}\theta_\alpha \,\mathrm{d}\theta_\beta}{\sum_{\alpha,\beta} a_{\alpha\beta} \,\mathrm{d}\theta_\alpha \,\mathrm{d}\theta_\beta}. \tag{1.19}$$

The quadratic form in the numerator on the right-hand side of equation (1.19) is called the *second fundamental form* of the surface, and we see that the normal curvature is determined by the ratio between the first and the second fundamental forms.

It appears from equation (1.14) and Fig. 3 that the radius of curvature R of the normal section is a signed quantity, R being positive when the vector $\overrightarrow{PQ}$ from the intersection P between normal and surface to the centre of curvature Q has the same direction as $\mathbf{a}_3$.

We recall that the *principal directions* at any point on the middle surface are two mutually perpendicular directions in the corresponding tangent plane with the property that the normal curvatures at that point assume extreme values in these two directions (the *principal curvatures*). At each point of the surface there is at least one set of principal directions. A *line of curvature* is a curve on the surface with the property that, at any point of the curve, the tangent is a principal direction of the surface at that point. Two lines of curvature intersecting at right-angles pass through each point of the surface.

1.2.1 Lines of curvature as coordinate curves

In the previous section we made no special assumptions about the curvilinear coordinates of the middle surface. However, in the following we shall assume that the coordinate curves are the *lines of curvature*. This system of coordinate curves has particularly simple properties, so that the equations of the theory of shells, when written in full, assume a relatively simple form in this system. On the other hand, the use of the special curvilinear coordinates implies a certain restriction on the field of application of the theory. In order to be able to use the following formulae, a knowledge of the lines of curvature is required, and the determination of these curves for a given surface is, in general, a fairly complicated problem. However, for many of the types of shells used in practice, the geometry of the middle surface is of a simple nature (e.g. surfaces of revolution or cylindrical surfaces), so that the lines of curvature are already known. In such cases the following formulae can be used directly.

As the coordinate curves are lines of curvature, the θ_1- and θ_2-curves are mutually orthogonal families of curves. It follows that

$$\mathbf{a}_1 \cdot \mathbf{a}_2 = 0. \tag{1.20}$$

The lengths of the base vectors are denoted by A_1 and A_2, and we have

$$A_\alpha = |\mathbf{a}_\alpha| = (\mathbf{a}_\alpha \cdot \mathbf{a}_\alpha)^{1/2}. \tag{1.21}$$

The quantities A_α ($\alpha = 1, 2$) are called the *Lamé parameters* of the surface. By using (1.8) and (1.21) we find for the lengths ds_α of line elements along the coordinate curves:

$$(ds_\alpha)^2 = \mathbf{a}_\alpha \cdot \mathbf{a}_\alpha \, d\theta_\alpha^2 = A_\alpha^2 \, d\theta_\alpha^2,$$

or

$$ds_\alpha = A_\alpha \, d\theta_\alpha, \tag{1.22}$$

in which it is assumed that the arc length s_α is an increasing function of θ_α.

The area dA of a small surface element bounded by coordinate curves and with side lengths ds_1 and ds_2 (see Fig. 4) is given by

$$dA = ds_1 \, ds_2 = G \, d\theta_1 \, d\theta_2, \tag{1.23}$$

where

$$G = A_1 A_2. \tag{1.23a}$$

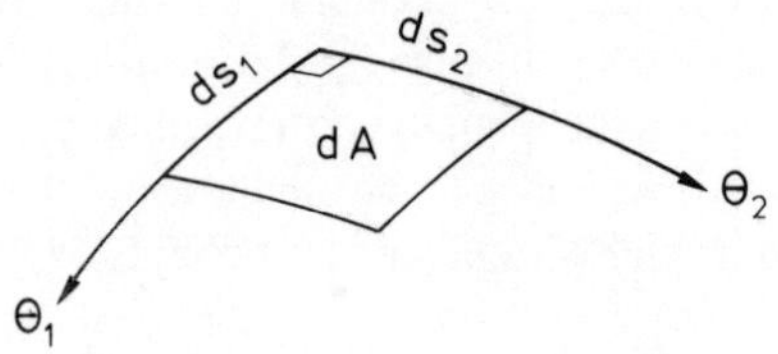

Fig. 4. Area of surface element

It is shown in the theory of surfaces that the necessary and sufficient condition that the coordinate curves be lines of curvature is that the equations

$$\mathbf{a}_1 \cdot \mathbf{a}_2 = 0, \qquad \mathbf{a}_3 \cdot \mathbf{a}_{1,2} = 0 \tag{1.24}$$

are satisfied everywhere on the surface. By the use of (1.9) and (1.18) it follows that these conditions can, alternatively, be written in the form

$$a_{12} = 0, \qquad b_{12} = 0. \tag{1.25}$$

Application of (1.13) and (1.5) gives

$$\mathbf{a}_3 \cdot \mathbf{a}_{1,2} = -\mathbf{a}_{3,2} \cdot \mathbf{a}_1,$$

$$\mathbf{a}_3 \cdot \mathbf{a}_{1,2} = \mathbf{a}_3 \cdot \mathbf{a}_{2,1} = -\mathbf{a}_{3,1} \cdot \mathbf{a}_2,$$

so that the second condition (1.24) is equivalent to the following two equations:

$$\mathbf{a}_{3,1} \cdot \mathbf{a}_2 = 0, \qquad \mathbf{a}_{3,2} \cdot \mathbf{a}_1 = 0. \tag{1.26}$$

These equations state that the derivative of the normal vector along one of the lines of curvature has no component in the direction of the tangent to the second line of curvature. The equations (1.26) are satisfied because we use the lines of curvature as coordinate curves (for other systems of coordinate curves these equations will not generally be true).

At each point of the middle surface we now introduce three mutually orthogonal unit vectors $\mathbf{e}_1$, $\mathbf{e}_2$, $\mathbf{e}_3$ in the directions of the base vectors and the normal, respectively. These unit vectors are therefore given by

$$\mathbf{e}_\alpha = \mathbf{a}_\alpha / A_\alpha \quad (\alpha = 1, 2), \qquad \mathbf{e}_3 = \mathbf{a}_3, \tag{1.27}$$

and the following relations are valid everywhere on the surface:

$$\begin{aligned} \mathbf{e}_\alpha \cdot \mathbf{e}_\alpha &= 1, \qquad & \mathbf{e}_3 \cdot \mathbf{e}_3 &= 1, \\ \mathbf{e}_1 \cdot \mathbf{e}_2 &= 0, \qquad & \mathbf{e}_3 \cdot \mathbf{e}_\alpha &= 0. \end{aligned} \tag{1.28}$$

We shall now determine the derivatives of these unit vectors with respect to θ_1 and θ_2. Differentiating the second equation (1.28) we get

$$\mathbf{e}_{3,1} \cdot \mathbf{e}_3 = 0, \tag{1.29}$$

and dividing the first equation (1.26) by A_2, we find that

$$\mathbf{e}_{3,1} \cdot \mathbf{e}_2 = 0. \tag{1.30}$$

Moreover, by using (1.14) with $\mathbf{t} = \mathbf{e}_1$, we obtain

$$\frac{1}{R_1} = -\frac{\mathrm{d}\mathbf{e}_3}{\mathrm{d}s_1} \cdot \mathbf{e}_1 = -\frac{1}{A_1} \mathbf{e}_{3,1} \cdot \mathbf{e}_1,$$

since $\mathrm{d}s_1 = A_1 \, \mathrm{d}\theta_1$ (see (1.22)). Hence we have

$$\mathbf{e}_{3,1} \cdot \mathbf{e}_1 = -A_1/R_1, \tag{1.31}$$

where $1/R_1$ is the *principal curvature* associated with the θ_1-direction. We have now determined the projections of the vector $\mathbf{e}_{3,1}$ on the three mutually orthogonal unit vectors $\mathbf{e}_1$, $\mathbf{e}_2$, $\mathbf{e}_3$, and the vector itself is therefore given by (see (1.29), (1.30), and (1.31))

$$\mathbf{e}_{3,1} = -\frac{A_1}{R_1} \mathbf{e}_1. \tag{1.32}$$

In a similar manner we find

$$\mathbf{e}_{3,2} = -\frac{A_2}{R_2} \mathbf{e}_2. \tag{1.32a}$$

Differentiating the equations $\mathbf{e}_1 \cdot \mathbf{e}_1 = 1$ and $\mathbf{e}_3 \cdot \mathbf{e}_1 = 0$, we get

$$\begin{aligned} &\mathbf{e}_{1,2} \cdot \mathbf{e}_1 = 0, \\ &\mathbf{e}_{1,2} \cdot \mathbf{e}_3 = -\mathbf{e}_1 \cdot \mathbf{e}_{3,2} = \mathbf{e}_1 \cdot \frac{A_2}{R_2} \mathbf{e}_2 = 0, \end{aligned} \tag{1.33}$$

where (1.32a) has been used in the last equation. We also have

$$\mathbf{e}_{1,2} \cdot \mathbf{e}_2 = (\mathbf{a}_1/A_1)_{,2} \cdot \mathbf{e}_2 = \left(\frac{\mathbf{a}_{1,2}}{A_1} - \frac{A_{1,2}}{A_1^2} \mathbf{a}_1 \right) \cdot \mathbf{e}_2.$$

It follows from (1.28) and (1.5) that $\mathbf{a}_1 \cdot \mathbf{e}_2 = A_1 \mathbf{e}_1 \cdot \mathbf{e}_2 = 0$ and $\mathbf{a}_{1,2} = \mathbf{a}_{2,1}$.

Thus

$$\begin{aligned} \mathbf{e}_{1,2}\cdot\mathbf{e}_2 &= \frac{1}{A_1}\,\mathbf{a}_{1,2}\cdot\mathbf{e}_2 = \frac{1}{A_1}\,\mathbf{a}_{2,1}\cdot\frac{\mathbf{a}_2}{A_2} = \frac{1}{A_1A_2}\,\tfrac{1}{2}(\mathbf{a}_2\cdot\mathbf{a}_2)_{,1} \\ &= \frac{1}{A_1A_2}\,\tfrac{1}{2}(A_2^2)_{,1} = \frac{A_{2,1}}{A_1}. \end{aligned} \tag{1.34}$$

Using (1.33) and (1.34), we then obtain

$$\mathbf{e}_{1,2} = \frac{A_{2,1}}{A_1}\,\mathbf{e}_2. \tag{1.35}$$

A similar equation can be derived for $\mathbf{e}_{2,1}$ (this equation is obtained from (1.35) by interchanging the indices 1 and 2). Differentiating the equations $\mathbf{e}_2\cdot\mathbf{e}_2 = 1$, $\mathbf{e}_2\cdot\mathbf{e}_3 = 0$, $\mathbf{e}_2\cdot\mathbf{e}_1 = 0$, we finally have

$$\begin{aligned} \mathbf{e}_{2,2}\cdot\mathbf{e}_2 &= 0, \\ \mathbf{e}_{2,2}\cdot\mathbf{e}_3 &= -\mathbf{e}_2\cdot\mathbf{e}_{3,2} = A_2/R_2, \\ \mathbf{e}_{2,2}\cdot\mathbf{e}_1 &= -\mathbf{e}_2\cdot\mathbf{e}_{1,2} = -A_{2,1}/A_1, \end{aligned}$$

in which we have used (1.32a) and (1.35). Hence

$$\mathbf{e}_{2,2} = -\frac{A_{2,1}}{A_1}\,\mathbf{e}_1 + \frac{A_2}{R_2}\,\mathbf{e}_3, \tag{1.36}$$

and a similar equation holds for $\mathbf{e}_{1,1}$. Collecting the previous results, we have (see (1.32), (1.35) and (1.36))

$$\begin{aligned} \mathbf{e}_{1,1} &= -\frac{A_{1,2}}{A_2}\,\mathbf{e}_2 + \frac{A_1}{R_1}\,\mathbf{e}_3, & \mathbf{e}_{1,2} &= \frac{A_{2,1}}{A_1}\,\mathbf{e}_2, \\ \mathbf{e}_{2,1} &= \frac{A_{1,2}}{A_2}\,\mathbf{e}_1, & \mathbf{e}_{2,2} &= -\frac{A_{2,1}}{A_1}\,\mathbf{e}_1 + \frac{A_2}{R_2}\,\mathbf{e}_3, \\ \mathbf{e}_{3,1} &= -\frac{A_1}{R_1}\,\mathbf{e}_1, & \mathbf{e}_{3,2} &= -\frac{A_2}{R_2}\,\mathbf{e}_2. \end{aligned} \tag{1.37}$$

We now consider a *vector field* on the middle surface which associates a vector with each point of the middle surface. In view of the one-to-one correspondence between points of the middle surface and curvilinear coordinates θ_α, the vector field can therefore be written in the form $\mathbf{V}(\theta_1, \theta_2)$. At each point of the middle surface we now resolve the vector $\mathbf{V}$ in the directions of the unit vectors $\mathbf{e}_1$, $\mathbf{e}_2$, $\mathbf{e}_3$, i.e. we put

$$\mathbf{V}(\theta_1, \theta_2) = V_1\mathbf{e}_1 + V_2\mathbf{e}_2 + V_3\mathbf{e}_3 = \sum_{i=1}^{3} V_i\mathbf{e}_i, \tag{1.38}$$

where we have introduced the convention that *latin subscripts* represent the numbers 1, 2, 3. Differentiating (1.38) with respect to θ_α we get

$$\mathbf{V}_{,\alpha} = \sum_{i=1}^{3} (V_{i,\alpha}\mathbf{e}_i + V_i\mathbf{e}_{i,\alpha}). \tag{1.39}$$

Substituting the expressions (1.37) for the derivatives $\mathbf{e}_{i,\alpha}$ in (1.39) we obtain the following important formulae, which express the derivatives of the vector $\mathbf{V}$ in terms of its components:

$$\mathbf{V}_{,1} = \left(V_{1,1} + \frac{A_{1,2}}{A_2}V_2 - \frac{A_1}{R_1}V_3\right)\mathbf{e}_1 + \left(V_{2,1} - \frac{A_{1,2}}{A_2}V_1\right)\mathbf{e}_2 + \left(V_{3,1} + \frac{A_1}{R_1}V_1\right)\mathbf{e}_3,$$

$$\mathbf{V}_{,2} = \left(V_{1,2} - \frac{A_{2,1}}{A_1}V_2\right)\mathbf{e}_1 + \left(V_{2,2} + \frac{A_{2,1}}{A_1}V_1 - \frac{A_2}{R_2}V_3\right)\mathbf{e}_2 + \left(V_{3,2} + \frac{A_2}{R_2}V_2\right)\mathbf{e}_3. \tag{1.40}$$

We finally note that it follows from equations (1.25), (1.21) and (1.32) that the coefficients of the first and second fundamental forms of the surface in the present special coordinate system are given by

$$\begin{aligned} a_{11} &= A_1^2, & a_{12} &= a_{21} = 0, & a_{22} &= A_2^2, \\ b_{11} &= \frac{A_1^2}{R_1}, & b_{12} &= b_{21} = 0, & b_{22} &= \frac{A_2^2}{R_2}. \end{aligned} \tag{1.41}$$

1.2.2 Equations of Codazzi and Gauss

We shall now derive some important relations between the Lamé parameters A_1 and A_2 and the principal curvatures $1/R_1$ and $1/R_2$.

The equation

$$(\mathbf{e}_{3,1})_{,2} = (\mathbf{e}_{3,2})_{,1},$$

is evidently true, since it merely states that the order of differentiation is immaterial. Substituting the expressions (1.37) for $\mathbf{e}_{3,1}$ and $\mathbf{e}_{3,2}$ we get

$$\left(\frac{A_1}{R_1}\mathbf{e}_1\right)_{,2} = \left(\frac{A_2}{R_2}\mathbf{e}_2\right)_{,1}. \tag{1.42}$$

We now use the formulae (1.40) to calculate the derivatives in (1.42) of the vectors $(A_1/R_1)\mathbf{e}_1$ and $(A_2/R_2)\mathbf{e}_2$ with components $(A_1/R_1, 0, 0)$ and $(0, A_2/R_2, 0)$, respectively. On calculating the 1- and 2-components of the vector equation (1.42), we obtain

$$\left(\frac{A_1}{R_1}\right)_{,2} = \frac{A_{1,2}}{R_2}, \qquad \left(\frac{A_2}{R_2}\right)_{,1} = \frac{A_{2,1}}{R_1}, \tag{1.43}$$

while the calculation of the 3-component merely provides the trivial result $0 = 0$. The two equations (1.43) are called *Codazzi's equations*. By performing the differentiation on the left-hand sides of the equations, (1.43) can, alternatively, be written in the form

$$\left(\frac{1}{R_1}\right)_{,2} = \frac{A_{1,2}}{A_1}\left(\frac{1}{R_2} - \frac{1}{R_1}\right), \qquad \left(\frac{1}{R_2}\right)_{,1} = \frac{A_{2,1}}{A_2}\left(\frac{1}{R_1} - \frac{1}{R_2}\right). \tag{1.44}$$

We now consider the equation $(\mathbf{e}_{1,1})_{,2} = (\mathbf{e}_{1,2})_{,1}$. Using (1.37), we obtain

$$\left(-\frac{A_{1,2}}{A_2}\mathbf{e}_2 + \frac{A_1}{R_1}\mathbf{e}_3\right)_{,2} = \left(\frac{A_{2,1}}{A_1}\mathbf{e}_2\right)_{,1} \tag{1.45}$$

The calculation of the 2-component of this vector equation by means of (1.40) gives the following result

$$\left(\frac{A_{2,1}}{A_1}\right)_{,1} + \left(\frac{A_{1,2}}{A_2}\right)_{,2} = -\frac{A_1A_2}{R_1R_2}. \tag{1.46}$$

This equation is called *Gauss' equation*. The remaining two components of (1.45) do not give rise to any new relations (one of them is trivial and the other is identical with the Codazzi equation $(1.43)_1$). Similarly, it is found that the equation $(\mathbf{e}_{2,1})_{,2} = (\mathbf{e}_{2,2})_{,1}$ does not lead to any new relations.

We have now shown that the Lamé parameters and principal curvatures of a given surface satisfy the equations of Gauss and Codazzi. Let us now suppose that four functions A_1, A_2, R_1 and R_2 of (θ_1, θ_2) are given, and let us seek to determine a surface for which these functions are the Lamé parameters and principal radii of curvature. If such a surface exists, then A_1, A_2, R_1 and R_2 will satisfy the equations of Gauss and Codazzi, so that these equations constitute necessary conditions for the existence of the surface. It can be shown that these conditions are also sufficient to ensure the existence of the surface, provided merely that $A_1 > 0$ and $A_2 > 0$. The geometry of the surface is then determined except for a displacement as a rigid body in space.

It is a condition for the validity of the above results that the coordinate curves are lines of curvature. It can be shown, however, that similar results hold in the general case of arbitrary curvilinear coordinates on the surface. In this case the equations of Gauss and Codazzi take the form of three equations involving the coefficients $a_{\alpha\beta}$ and $b_{\alpha\beta}$ of the first and second fundamental forms of the surface (note that there are only six independent coefficients because of the symmetry conditions $a_{\alpha\beta} = a_{\beta\alpha}$ and $b_{\alpha\beta} = b_{\beta\alpha}$), these three equations reducing to (1.43) and (1.46) when the coordinate curves are lines of curvature (see section 4.3). Corresponding to the above result, the following *fundamental theorem* is valid in the present case: The necessary and sufficient conditions that a surface exists, for which the coefficients $a_{\alpha\beta}$ and $b_{\alpha\beta}$ of the first and second fundamental forms are six given functions, are that these functions satisfy the equations of Gauss and Codazzi, and that the quadratic form $\sum_{\alpha,\beta} a_{\alpha\beta}\,\mathrm{d}\theta_\alpha\,\mathrm{d}\theta_\beta$ is everywhere positive definite. When these conditions are satisfied, the geometry of the surface is determined except for a displacement as a rigid body in space.

We finally mention that the quantity

$$K = \frac{1}{R_1R_2} \tag{1.47}$$

is called the *Gaussian curvature*. A point of the surface is called elliptic, parabolic or hyperbolic according to whether K is positive, zero or negative. It can be shown that if K is zero at all points of the surface, then the surface is *developable*.

1.3 General principles in the theory of shells

For the purpose of analysis, a shell may be regarded as a three-dimensional body, and the methods of continuum mechanics may then be applied (e.g. the theory of elasticity of three-dimensional bodies). However, a calculation based on these principles will generally be very difficult and complicated.

In the *theory of shells*, an alternative simplified method is therefore employed, in which the shell is regarded as a *two-dimensional body*. This two-dimensional theory is based on the fundamental assumption that the geometrical and statical quantities which determine the behaviour of the shell can be assumed with sufficient accuracy to be functions of only two independent variables (namely the curvilinear coordinates of the middle surface).

The development of the theory of shells can now proceed in different ways:

(1) The two-dimensional theory may be regarded as an approximation to the three-dimensional theory. In this case the two-dimensional theory can be derived either (a) as a limiting case of the three-dimensional theory which occurs when the thickness of the shell becomes very small, or (b) by introducing special assumptions concerning the displacements of the three-dimensional shell, namely that the displacements throughout the shell are determined by the displacements of the middle surface. However, certain difficulties are encountered when the two-dimensional theory is derived from the three-dimensional theory. For instance, the displacement field given by the exact solution according to the three-dimensional theory will generally be different from the displacement field assumed in the above method (b), and this circumstance gives rise to certain contradictions.

(2) Alternatively, the theory of shells may be developed as a purely two-dimensional theory without any reference to the three-dimensional theory. In this way, the difficulties which arise when we seek to interrelate the two- and three-dimensional theories are avoided. A derivation of the theory of shells as a purely two-dimensional theory is based on unambiguous assumptions which do not lead to contradictions. Because of these advantages of the two-dimensional approach, we shall use this method in the following derivation of the theory of shells.

We shall therefore regard the shell as a *two-dimensional body*. By this we mean a collection of material points situated on a surface (the middle surface). When the shell is in motion, the positions of the material points (and thus, in general, the shape of the surface) vary in time. It is assumed that, at any moment of time, the material points occupy a region on the surface so that there is a one-to-one correspondence between the material points and the points of the surface in this region. The positions in space of the material points determine the *configuration* of the shell at that moment.

1.4 Geometrical description

1.4.1 The displacements of the middle surface

We consider a displacement of the two-dimensional body which carries it from a configuration called the *reference state* into another configuration, which is called the *deformed state*.

We introduce curvilinear coordinates (θ_1, θ_2) on the middle surface in the reference state by using the lines of curvature as coordinate curves. All the geometrical results derived in sections 1.1 and 1.2 are therefore valid for the middle surface in the reference state.

Consider a material point of the middle surface. As a result of the displacement, this material point moves from a point P on the middle surface in the reference state to a point P_1 on the middle surface in the deformed state (see Fig. 5). The displacement will depend on the chosen material point.

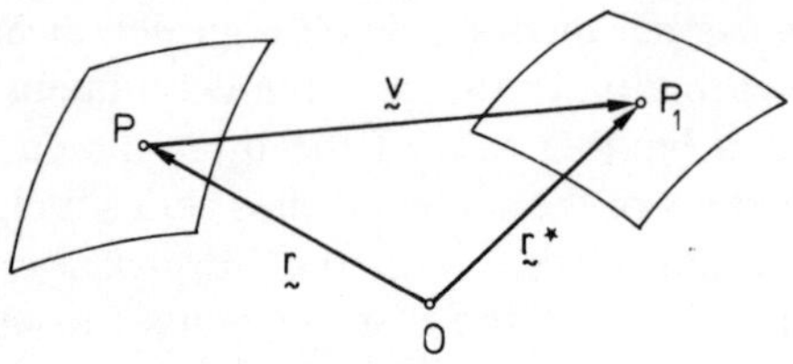

Fig. 5. Displacement vector

Identifying the material point by its position on the middle surface in the reference state, i.e. by means of the curvilinear coordinates (θ_1, θ_2) of the point P, we can therefore write the displacement vector in the form $\mathbf{v}(\theta_1, \theta_2)$. Let $\mathbf{r}^*$ denote the position vector of the material point in the deformed state, i.e. of the point P_1. From the vector equation $\overrightarrow{OP_1} = \overrightarrow{OP} + \overrightarrow{PP_1}$ (see Fig. 5) we then obtain the following expression for $\mathbf{r}^*$:

$$\mathbf{r}^* = \mathbf{r}(\theta_1, \theta_2) + \mathbf{v}(\theta_1, \theta_2). \tag{1.48}$$

This is the parametric representation of the *middle surface* in the *deformed state*. Since a material point has the same curvilinear coordinates in the reference state and in the deformed state, the coordinates (θ_1, θ_2) are *convected coordinates*. Material points situated on a coordinate curve in the reference state are therefore carried into a curve which is a coordinate curve on the middle surface in the deformed state. It should be noted that the coordinate curves of the middle surface in the deformed state will *not generally be lines of curvature* for that surface.

Corresponding to the geometrical quantities introduced in section 1.2, we can now define two sets of quantities, one set for the reference state and one

Table 1.1 Quantities in reference and deformed states

	Reference state	Deformed state
Position vector	$\mathbf{r}(\theta_1, \theta_2)$	$\mathbf{r}^*(\theta_1, \theta_2)$
Base vectors	$\mathbf{a}_\alpha = \partial\mathbf{r}/\partial\theta_\alpha$	$\mathbf{a}_\alpha^* = \partial\mathbf{r}^*/\partial\theta_\alpha$
Unit normal vector	$\mathbf{a}_3$	$\mathbf{a}_3^*$
Lamé parameters	$A_\alpha = (\mathbf{a}_\alpha \cdot \mathbf{a}_\alpha)^{1/2}$	$A_\alpha^* = (\mathbf{a}_\alpha^* \cdot \mathbf{a}_\alpha^*)^{1/2}$
Coefft of 1st fund. form	$a_{\alpha\beta} = \mathbf{a}_\alpha \cdot \mathbf{a}_\beta$	$a_{\alpha\beta}^* = \mathbf{a}_\alpha^* \cdot \mathbf{a}_\beta^*$
Coefft of 2nd fund. form	$b_{\alpha\beta} = \mathbf{a}_{\alpha,\beta} \cdot \mathbf{a}_3$	$b_{\alpha\beta}^* = \mathbf{a}_{\alpha,\beta}^* \cdot \mathbf{a}_3^*$

set for the deformed state. We shall denote quantities belonging to the deformed state by *asterisks* (see Table 1.1).

Infinitesimal theory. In the following derivation of the theory of shells we shall assume that the *displacements* and *rotations* of the middle surface are *infinitesimally small*.

All the quantities in the above table are functions of the position vector of the middle surface and its partial derivatives. Each of the quantities belonging to the deformed state can be obtained from the formula that defines the quantity in the reference state by replacing the position vector **r** (reference state) by the position vector $\mathbf{r}^* = \mathbf{r} + \mathbf{v}$ (deformed state) in this formula.

Let us consider one of these quantities, which is a function of the partial derivatives of the position vector of the first, second and up to the pth order, i.e.

$$f(\mathbf{r}_{,1}, \mathbf{r}_{,2}, \mathbf{r}_{,11}, \mathbf{r}_{,22}, \mathbf{r}_{,12}, \ldots) = f(\mathbf{R}),$$

where **R** denotes a vector, the components of which are the arguments of the function f, i.e. the components of all the vectors between the parentheses on the left-hand side of the equation. We shall assume that f is a differentiable function, i.e. that the increment of the function for infinitesimal increments of the arguments is linear in these.

We can find the value of this quantity in the deformed state by replacing the position vector **r** by the position vector $\mathbf{r}^* = \mathbf{r} + \mathbf{v}$ so that we have

$$f^* = f(\mathbf{r}_{,1} + \mathbf{v}_{,1}, \mathbf{r}_{,2} + \mathbf{v}_{,2}, \mathbf{r}_{,11} + \mathbf{v}_{,11}, \ldots) = f(\mathbf{R} + \mathbf{V}),$$

where the components of all the vectors $\mathbf{v}_{,1}, \mathbf{v}_{,2}, \mathbf{v}_{,11}, \ldots$ have been collected to form the components of the single vector **V**. We shall now study the increment

$$f^* - f = f(\mathbf{R} + \mathbf{V}) - f(\mathbf{R}), \tag{1.49}$$

corresponding to the transition from the reference state to the deformed state. Because of the assumed infinitesimal displacements and rotations, we shall linearize the increment $(f^* - f)$ with respect to the displacement quantities, i.e. the function $(f^* - f)$, which is generally nonlinear in the displacement quantities, will be approximated by a function that is linear in these quantities.

The first variation. The linearized increment of the function is denoted by δf so that

$$(f^* - f)_{\text{lin}} = [f(\mathbf{R} + \mathbf{V}) - f(\mathbf{R})]_{\text{lin}} = \delta f. \tag{1.50}$$

δf is called the *first variation* of f and is linear in the components of $\mathbf{V}$. We shall now derive a general formula for the first variation. We consider the function

$$f(\mathbf{R} + t\mathbf{V}),$$

where t is a parameter. This is a function of θ_1, θ_2 and t, since $\mathbf{R}$ and $\mathbf{V}$ are functions of (θ_1, θ_2). We now form the partial derivative of this function with respect to t and then put $t = 0$. We shall show that the quantity obtained in this way is the first variation, i.e.

$$\delta f = \frac{\mathrm{d}}{\mathrm{d}t} f(\mathbf{R} + t\mathbf{V})_{|t=0}. \tag{1.51}$$

The proof will be confined to the simple case in which the vector $\mathbf{R}$ has only two components, so that

$$f = f(R_1(\theta_1, \theta_2), R_2(\theta_1, \theta_2)).$$

In accordance with the above rule, we form the quantity

$$f(R_1 + tV_1, R_2 + tV_2),$$

where V_1 and V_2 are functions of (θ_1, θ_2). Differentiating with respect to t, we derive

$$\frac{\mathrm{d}f}{\mathrm{d}t}_{|t=0} = \frac{\partial f}{\partial R_1} V_1 + \frac{\partial f}{\partial R_2} V_2, \tag{1.51a}$$

where the partial derivatives on the right-hand side of the equation are given by

$$\frac{\partial}{\partial R_\alpha} f(R_1, R_2), \qquad \alpha = 1, 2,$$

corresponding to $t = 0$, and R_α and V_α are functions of (θ_1, θ_2). It will now be seen that the right-hand side of (1.51a) is the linearized increment of the function, i.e. the first variation δf. This completes the proof for the case of a vector $\mathbf{R}$ with two components, but the proof can immediately be generalized to cases in which the number of components is greater than two.

The formula (1.51) will now be used to determine the first variation of the sum and the product of two functions f and g. We have

$$\delta(f + g) = \frac{\mathrm{d}}{\mathrm{d}t}[f(\mathbf{R} + t\mathbf{V}) + g(\mathbf{R} + t\mathbf{V})]_{|t=0} = \left(\frac{\mathrm{d}f}{\mathrm{d}t} + \frac{\mathrm{d}g}{\mathrm{d}t}\right)_{|t=0},$$

$$\delta(fg) = \frac{\mathrm{d}}{\mathrm{d}t}[f(\mathbf{R} + t\mathbf{V}) \cdot g(\mathbf{R} + t\mathbf{V})]_{|t=0} = \left(\frac{\mathrm{d}f}{\mathrm{d}t} g + f\frac{\mathrm{d}g}{\mathrm{d}t}\right)_{|t=0},$$

from which we deduce by means of (1.51) that

$$\delta(f+g) = \delta f + \delta g, \qquad \delta(fg) = \delta f\, g + f\, \delta g. \tag{1.52}$$

We note the analogy to the rules for the evaluation of differentials.

We further consider the first variation of the partial derivatives $f_{,\alpha} = \partial f/\partial\theta_\alpha$, and we recall that both **R** and **V** are functions of (θ_1, θ_2). Since t and θ_α are independent variables, so that the order of differentiation is immaterial, we find

$$\delta(f_{,\alpha}) = \left[\frac{\mathrm{d}}{\mathrm{d}t}\left(\frac{\partial f}{\partial\theta_\alpha}\right)\right]_{|t=0} = \left[\frac{\partial}{\partial\theta_\alpha}\left(\frac{\mathrm{d}f}{\mathrm{d}t}\right)\right]_{|t=0} = \frac{\partial}{\partial\theta_\alpha}\left(\frac{\mathrm{d}f}{\mathrm{d}t}_{|t=0}\right)$$

or, by means of (1.51),

$$\delta(f_{,\alpha}) = (\delta f)_{,\alpha}. \tag{1.53}$$

We have thus deduced the important result that *the order of the operations* $\delta(\ )$ *and* $(\ )_{,\alpha}$ *is immaterial.*

We shall now determine the first variation of the base vectors and the coefficients of the first fundamental form. On differentiating (1.48) we find, for the base vectors in the deformed state,

$$\mathbf{a}^*_\alpha = \mathbf{r}^*_{,\alpha} = \mathbf{r}_{,\alpha} + \mathbf{v}_{,\alpha},$$

or

$$\mathbf{a}^*_\alpha - \mathbf{a}_\alpha = \mathbf{v}_{,\alpha}. \tag{1.54}$$

As this increment is linear in $\mathbf{v}_{,\alpha}$, i.e. in the displacement derivatives, we deduce from (1.54) that

$$\delta\mathbf{a}_\alpha = (\mathbf{a}^*_\alpha - \mathbf{a}_\alpha)_{\text{lin}} = \mathbf{a}^*_\alpha - \mathbf{a}_\alpha = \mathbf{v}_{,\alpha}, \tag{1.54a}$$

so that

$$\delta\mathbf{a}_\alpha = \mathbf{v}_{,\alpha}. \tag{1.55}$$

The same result is, of course, obtained by means of (1.51).

The coefficients of the first fundamental form are defined by (see (1.9))

$$a_{\alpha\beta} = \mathbf{a}_\alpha \cdot \mathbf{a}_\beta.$$

With the help of (1.52) and (1.55) we therefore obtain

$$\delta(\mathbf{a}_\alpha \cdot \mathbf{a}_\beta) = \mathbf{a}_\alpha \cdot \delta\mathbf{a}_\beta + \mathbf{a}_\beta \cdot \delta\mathbf{a}_\alpha, \tag{1.56}$$

$$\delta a_{\alpha\beta} = (a^*_{\alpha\beta} - a_{\alpha\beta})_{\text{lin}} = \mathbf{a}_\alpha \cdot \mathbf{v}_{,\beta} + \mathbf{a}_\beta \cdot \mathbf{v}_{,\alpha}. \tag{1.57}$$

Consider an arbitrary displacement of the middle surface. At each point of the deformed middle surface the normal vector is orthogonal to the two base vectors. The equation

$$\mathbf{a}^*_\alpha \cdot \mathbf{a}^*_3 = 0$$

is therefore true for all displacements and, in particular, for displacements of the form $t\mathbf{v}$, where the value of t is arbitrary. For such a displacement, we therefore have

$$\frac{\mathrm{d}}{\mathrm{d}t}(\mathbf{a}_\alpha^* \cdot \mathbf{a}_3^*) = 0$$

for all t, and in particular for $t = 0$. We therefore obtain by means of (1.51)

$$\delta(\mathbf{a}_\alpha \cdot \mathbf{a}_3) = 0,$$

or, from $(1.52)_2$ and (1.55),

$$\mathbf{a}_\alpha \cdot \delta\mathbf{a}_3 = -\mathbf{v}_{,\alpha} \cdot \mathbf{a}_3. \tag{1.58}$$

In a similar manner, it follows from the equations (see (1.28))

$$\mathbf{e}_i^* \cdot \mathbf{e}_i^* = 1,$$

which are valid for all displacements, that

$$\delta(\mathbf{e}_i \cdot \mathbf{e}_i) = 0$$

or

$$\mathbf{e}_i \cdot \delta\mathbf{e}_i = 0, \tag{1.59}$$

where the index i can assume the values 1, 2, 3. This equation expresses the fact that the linearized increment of the unit vector $\mathbf{e}_i$ is perpendicular to this vector.

1.4.2 The middle surface strains

A displacement as a rigid body of the shell is one by which the length of every line element of the middle surface and the angle between the surface normals at every pair of material points are left unchanged. If these quantities change as a result of the displacement, the shell is said to be strained or deformed. In the following we shall develop a quantitative description of the deformation.

Let us consider a material line element of the middle surface, which joins the points with curvilinear coordinates θ_α and $(\theta_\alpha + \mathrm{d}\theta_\alpha)$. Let $\mathrm{d}s$ and $\mathrm{d}s^*$ be the lengths of the line element in the reference state and the deformed state, respectively. We now form the quantity

$$(\mathrm{d}s^*)^2 - (\mathrm{d}s)^2 = \sum_{\alpha,\beta} (a_{\alpha\beta}^* - a_{\alpha\beta})\,\mathrm{d}\theta_\alpha\,\mathrm{d}\theta_\beta, \tag{1.60}$$

where we have used (1.10) and the corresponding equation for the deformed state. If $a_{\alpha\beta}^* - a_{\alpha\beta} = 0$ everywhere on the middle surface, the lengths of all line elements remain unchanged, and the displacement is said to be *inextensional*. If the quantities $(a_{\alpha\beta}^* - a_{\alpha\beta})$ do not vanish identically, the lengths of line elements will change, and the middle surface will therefore be strained. It will be seen from the following that the quantities $(a_{\alpha\beta} - a_{\alpha\beta})$ are convenient for describing the middle surface strains.

Since the displacements are assumed to be infinitesimal, we shall linearize the quantities $(a^*_{\alpha\beta} - a_{\alpha\beta})$. The *strain measures* $\varepsilon_{\alpha\beta}$ of the middle surface are then defined in the following manner:

$$\varepsilon_{\alpha\beta} = \frac{1}{2A_\alpha A_\beta}(a^*_{\alpha\beta} - a_{\alpha\beta})_{\text{lin}} = \frac{1}{A_\alpha A_\beta}\tfrac{1}{2}\delta a_{\alpha\beta}. \tag{1.61}$$

Using (1.9) and (1.57), we find

$$\varepsilon_{\alpha\beta} = \frac{1}{A_\alpha A_\beta}\tfrac{1}{2}\delta(\mathbf{a}_\alpha \cdot \mathbf{a}_\beta) = \frac{1}{2A_\alpha A_\beta}(\mathbf{a}_\alpha \cdot \mathbf{v}_{,\beta} + \mathbf{a}_\beta \cdot \mathbf{v}_{,\alpha}). \tag{1.62}$$

It follows from this equation that the strain measures are symmetric in the indices α and β, i.e.

$$\varepsilon_{\alpha\beta} = \varepsilon_{\beta\alpha}. \tag{1.63}$$

From (1.62) we derive, with the help of (1.27),

$$\begin{aligned}
\varepsilon_{11} &= \frac{1}{A_1}\mathbf{e}_1 \cdot \mathbf{v}_{,1},\\
\varepsilon_{22} &= \frac{1}{A_2}\mathbf{e}_2 \cdot \mathbf{v}_{,2},\\
\varepsilon_{12} = \varepsilon_{21} &= \frac{1}{2}\left(\frac{\mathbf{e}_1 \cdot \mathbf{v}_{,2}}{A_2} + \frac{\mathbf{e}_2 \cdot \mathbf{v}_{,1}}{A_1}\right).
\end{aligned} \tag{1.64}$$

We now resolve the displacement vector $\mathbf{v}$ in the directions of the unit vectors $\mathbf{e}_i$ (see (1.27)). Hence

$$\mathbf{v} = \sum_{i=1}^{3} v_i \mathbf{e}_i. \tag{1.65}$$

Using the equations (1.40) for the derivatives of a vector, we find from (1.64) that

$$\begin{aligned}
\varepsilon_{11} &= \frac{1}{A_1}v_{1,1} + \frac{A_{1,2}}{A_1A_2}v_2 - \frac{v_3}{R_1},\\
\varepsilon_{12} = \varepsilon_{21} &= \frac{1}{2}\left(\frac{1}{A_2}v_{1,2} + \frac{1}{A_1}v_{2,1} - \frac{A_{1,2}}{A_1A_2}v_1 - \frac{A_{2,1}}{A_1A_2}v_2\right)\\
&= \frac{1}{2}\left[\frac{A_1}{A_2}\left(\frac{v_1}{A_1}\right)_{,2} + \frac{A_2}{A_1}\left(\frac{v_2}{A_2}\right)_{,1}\right],\\
\varepsilon_{22} &= \frac{1}{A_2}v_{2,2} + \frac{A_{2,1}}{A_1A_2}v_1 - \frac{v_3}{R_2}.
\end{aligned} \tag{1.66}$$

These formulae express the *strain measures of the middle surface* in terms of the displacement components.

A geometrical interpretation of the strain measures is provided by the following considerations. From equations (1.9) and (1.27) we obtain

$$a_{11} = \mathbf{a}_1 \cdot \mathbf{a}_1 = A_1\mathbf{e}_1 \cdot A_1\mathbf{e} = A_1^2,$$

which, on substitution in (1.61), yields

$$\varepsilon_{11} = \frac{1}{A_1^2}\tfrac{1}{2}\delta(A_1^2) = \frac{\delta A_1}{A_1}. \tag{1.67}$$

The strain measure ε_{11} is therefore the elongation per unit length (the extension) in the direction of the θ_1-curve. In a similar manner, it is found that ε_{22} is the elongation per unit length in the direction of the θ_2-curve. Using (1.27), (1.52) and (1.55), we further derive

$$\mathbf{e}_1 \cdot \delta\mathbf{e}_2 = \mathbf{e}_1 \cdot \delta\left(\frac{\mathbf{a}_2}{A_2}\right) = \mathbf{e}_1 \cdot \left(\frac{\delta\mathbf{a}_2}{A_2} - \frac{\delta A_2}{A_2^2}\mathbf{a}_2\right) = \mathbf{e}_1 \cdot \frac{\mathbf{v}_{,2}}{A_2},$$

since $\mathbf{e}_1 \cdot \mathbf{a}_2 = \mathbf{e}_1 \cdot A_2\mathbf{e}_2 = 0$. In a similar manner, we find the relation

$$\mathbf{e}_2 \cdot \delta\mathbf{e}_1 = \mathbf{e}_2 \cdot \mathbf{v}_{,1}/A_1.$$

Inserting these results in $(1.64)_3$, we obtain

$$\varepsilon_{12} = \varepsilon_{21} = \tfrac{1}{2}(\mathbf{e}_1 \cdot \delta\mathbf{e}_2 + \mathbf{e}_2 \cdot \delta\mathbf{e}_1). \tag{1.68}$$

We now recall that $\delta\mathbf{e}_\alpha$ is orthogonal to the unit vector $\mathbf{e}_\alpha$ (see (1.59)). It can therefore be seen from (1.68) and Fig. 6 that $2\varepsilon_{12}$ is the decrease of the

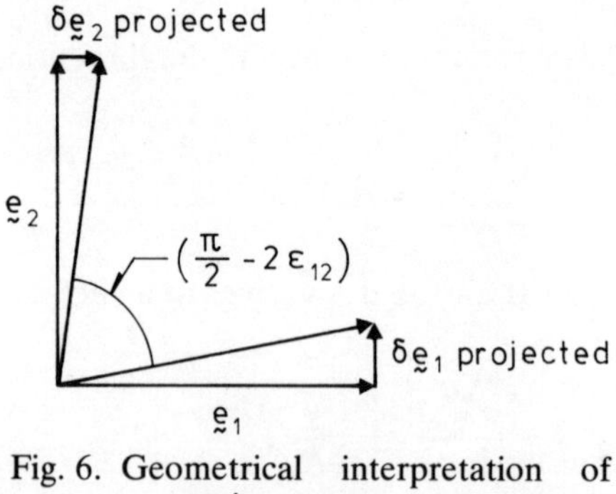

Fig. 6. Geometrical interpretation of strain measure ε_{12}

originally right angle between material line elements in the θ_1- and θ_2-directions. It follows that our definition of strain for the shell agrees with the usual definition of strain in the linear, infinitesimal theory of plane strain.

1.4.3 The rotation vector

The increments of the base vectors corresponding to the infinitesimal displacement are given by (see (1.54a))

$$\mathbf{a}_\alpha^* - \mathbf{a}_\alpha = \delta\mathbf{a}_\alpha.$$

We now write these increments as a sum of two contributions, i.e.

$$\delta \mathbf{a}_\alpha = \delta \mathbf{a}_\alpha^{(1)} + \delta \mathbf{a}_\alpha^{(2)}. \tag{1.69}$$

For the first contribution, we choose the expression

$$\delta \mathbf{a}_\alpha^{(1)} = A_\alpha \sum_\beta \varepsilon_{\alpha\beta} \mathbf{e}_\beta, \tag{1.70}$$

or, in full,

$$\begin{aligned} \delta \mathbf{a}_1^{(1)} &= A_1(\varepsilon_{11}\mathbf{e}_1 + \varepsilon_{12}\mathbf{e}_2), \\ \delta \mathbf{a}_2^{(1)} &= A_2(\varepsilon_{21}\mathbf{e}_1 + \varepsilon_{22}\mathbf{e}_2). \end{aligned} \tag{1.71}$$

Let us consider a point P of the middle surface in the reference state, and the tangent plane through P. We shall regard the associated base vectors $\mathbf{a}_\alpha$ as localized vectors in the tangent plane, whose initial points coincide with P. Equations (1.71) show that the vectors $\delta \mathbf{a}_\alpha^{(1)}$ are parallel to the above tangent plane. If we therefore add $\delta \mathbf{a}_\alpha^{(1)}$ to $\mathbf{a}_\alpha$, the sum $\mathbf{a}_\alpha + \delta \mathbf{a}_\alpha^{(1)}$ will lie on the tangent plane. Fig. 7 shows this tangent plane with the base vectors $\mathbf{a}_\alpha$ and the increments $\delta \mathbf{a}_\alpha^{(1)}$. It can be seen from the figure that the length of the vector

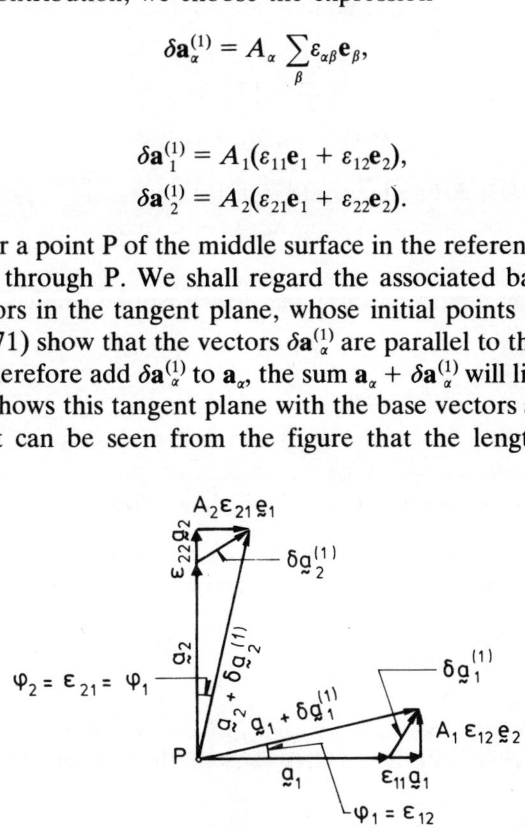

Fig. 7. Increments of base vectors

$\mathbf{a}_1 + \delta \mathbf{a}_1^{(1)}$ is equal to $(1 + \varepsilon_{11})|\mathbf{a}_1| = (1 + \varepsilon_{11})A_1$ (since the displacements are infinitesimal), and in a similar way it is found that the length of the vector $\mathbf{a}_2 + \delta \mathbf{a}_2^{(1)}$ is equal to $(1 + \varepsilon_{22})A_2$. It will further be seen that the angle between the vectors $\mathbf{a}_1 + \delta \mathbf{a}_1^{(1)}$ and $\mathbf{a}_2 + \delta \mathbf{a}_2^{(1)}$ is $\pi/2 - 2\varepsilon_{12}$. Now it follows from the discussion of the geometric meaning of the strain measures in section 1.4.2 that the length of the base vector $\mathbf{a}_\alpha^*$ is $(1 + \varepsilon_{\alpha\alpha})A_\alpha$, and that the angle between the base vectors $\mathbf{a}_1^*$ and $\mathbf{a}_2^*$ is $\pi/2 - 2\varepsilon_{12}$, i.e. the same values as those associated with the vectors $\mathbf{a}_\alpha + \delta \mathbf{a}_\alpha^{(1)}$. We conclude that the vectors $\mathbf{a}_\alpha + \delta \mathbf{a}_\alpha^{(1)}$ can be carried into the vectors $\mathbf{a}_\alpha^*$ by an infinitesimal rotation. The second contribution $\delta \mathbf{a}_\alpha^{(2)}$ in (1.69) therefore corresponds to an infinitesimal rotation, which can be represented by a *rotation vector* $\boldsymbol{\omega}$. We therefore have

$$\delta \mathbf{a}_\alpha^{(2)} = [\boldsymbol{\omega} \times (\mathbf{a}_\alpha + \delta \mathbf{a}_\alpha^{(1)})]_{\text{lin}} = \boldsymbol{\omega} \times \mathbf{a}_\alpha, \tag{1.72}$$

since the term $\boldsymbol{\omega} \times \delta\mathbf{a}_\alpha^{(1)}$ is omitted in the present infinitesimal theory (a product of small quantities). As the unit vector $\mathbf{a}_3$ is perpendicular to the vectors $\mathbf{a}_\alpha + \delta\mathbf{a}_\alpha^{(1)}$ and the unit vector $\mathbf{a}_3^*$ is perpendicular to the vectors $\mathbf{a}_\alpha^*$, it follows that $\mathbf{a}_3$ will be carried into $\mathbf{a}_3^*$ by the above rotation. The results of our considerations may now be summarized in the following equations (see (1.69), (1.70) and (1.72)):

$$\delta\mathbf{a}_\alpha = A_\alpha \sum_\beta \varepsilon_{\alpha\beta}\mathbf{e}_\beta + \boldsymbol{\omega} \times \mathbf{a}_\alpha, \tag{1.73}$$

$$\delta\mathbf{a}_3 = (\mathbf{a}_3^* - \mathbf{a}_3)_{\text{lin}} = \boldsymbol{\omega} \times \mathbf{a}_3. \tag{1.74}$$

Using (1.55) we can write (1.73) in the form

$$\mathbf{v}_{,\alpha} - \boldsymbol{\omega} \times \mathbf{a}_\alpha = A_\alpha \sum_\beta \varepsilon_{\alpha\beta}\mathbf{e}_\beta. \tag{1.75}$$

We also record the following formulae, which will be needed later:

$$(\mathbf{v}_{,\alpha} - \boldsymbol{\omega} \times \mathbf{a}_\alpha) \cdot \mathbf{a}_\beta = A_\alpha A_\beta \varepsilon_{\alpha\beta}, \tag{1.76}$$

$$(\mathbf{v}_{,\alpha} - \boldsymbol{\omega} \times \mathbf{a}_\alpha) \cdot \mathbf{a}_3 = 0. \tag{1.77}$$

These are derived by taking the scalar product of (1.75) with the vectors $\mathbf{a}_\beta$ and $\mathbf{a}_3$, respectively.

The components of the rotation vector. If we form the scalar product of (1.74) with $\mathbf{a}_\alpha$, we obtain

$$\mathbf{a}_\alpha \cdot \delta\mathbf{a}_3 = \mathbf{a}_\alpha \cdot (\boldsymbol{\omega} \times \mathbf{a}_3) = \boldsymbol{\omega} \cdot (\mathbf{a}_3 \times \mathbf{a}_\alpha) = \boldsymbol{\omega} \cdot \hat{\mathbf{a}}_\alpha. \tag{1.78}$$

The vector

$$\hat{\mathbf{a}}_\alpha = \mathbf{a}_3 \times \mathbf{a}_\alpha \tag{1.79}$$

is the *transverse vector* associated with $\mathbf{a}_\alpha$, i.e. the vector obtained by rotating $\mathbf{a}_\alpha$ through an angle $\pi/2$ in the positive sense in the tangent plane (see Fig. 8).

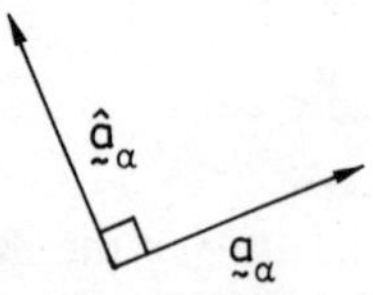

Fig. 8. Transverse vector

We therefore have the relations

$$\hat{\mathbf{e}}_1 = \mathbf{e}_2, \qquad \hat{\mathbf{e}}_2 = -\mathbf{e}_1. \tag{1.79a}$$

We now resolve the rotation vector $\boldsymbol{\omega}$ into components in the directions of the transverse vectors $\hat{\mathbf{e}}_\alpha$ and the normal vector $\mathbf{e}_3$, i.e.

$$\boldsymbol{\omega} = \omega_1\hat{\mathbf{e}}_1 + \omega_2\hat{\mathbf{e}}_2 + \omega_3\mathbf{e}_3. \tag{1.80}$$

Using (1.58) in (1.78), we derive

$$\boldsymbol{\omega} \cdot \hat{\mathbf{a}}_\alpha = -\mathbf{a}_3 \cdot \mathbf{v}_{,\alpha} \tag{1.81}$$

or

$$A_1\boldsymbol{\omega}\cdot\hat{\mathbf{e}}_1 = A_1\omega_1 = -\mathbf{a}_3\cdot\mathbf{v}_{,1},$$
$$A_2\boldsymbol{\omega}\cdot\hat{\mathbf{e}}_2 = A_2\omega_2 = -\mathbf{a}_3\cdot\mathbf{v}_{,2}. \tag{1.81a}$$

We now form two equations by writing the formula (1.76) for $(\alpha, \beta) = (1, 2)$ and $(\alpha, \beta) = (2, 1)$, respectively. Subtracting one of these equations from the other we obtain, since $\varepsilon_{12} = \varepsilon_{21}$,

$$\mathbf{a}_2\cdot\mathbf{v}_{,1} - \mathbf{a}_1\cdot\mathbf{v}_{,2} = 2\boldsymbol{\omega}\cdot(\mathbf{a}_1\times\mathbf{a}_2) = 2A_1A_2\boldsymbol{\omega}\cdot\mathbf{a}_3,$$

where we have used (1.11). It follows that

$$\omega_3 = \frac{1}{2A_1A_2}(\mathbf{a}_2\cdot\mathbf{v}_{,1} - \mathbf{a}_1\cdot\mathbf{v}_{,2}). \tag{1.82}$$

With the help of the formulae (1.40) for the derivatives of $\mathbf{v}$, we then obtain from (1.81a) and (1.82):

$$\omega_1 = -\left(\frac{1}{A_1}v_{3,1} + \frac{v_1}{R_1}\right),$$

$$\omega_2 = -\left(\frac{1}{A_2}v_{3,2} + \frac{v_2}{R_2}\right),$$

$$\omega_3 = \frac{1}{2A_1A_2}(A_2v_{2,1} - A_{1,2}v_1 - A_1v_{1,2} + A_{2,1}v_2) \tag{1.83}$$

$$= \frac{1}{2A_1A_2}[(A_2v_2)_{,1} - (A_1v_1)_{,2}].$$

It should be noted that the decomposition of $\delta\mathbf{a}_\alpha$ in the two contributions (1.70) and (1.72) corresponds to the unique representation of the infinitesimal displacement of a neighbourhood of a particle on the middle surface as the sum of a pure deformation and a displacement as a rigid body. The contribution $\delta\mathbf{a}_\alpha^{(1)}$ vanishes for any displacement as a rigid body (since the quantities $\varepsilon_{\alpha\beta}$ vanish in such displacements), and since the angles φ_1 and φ_2 in Fig. 7 are equal, it follows from known results (plane deformation) that the displacement corresponding to $\delta\mathbf{a}_\alpha^{(1)}$ is a pure deformation which gives rise to extension without rotation of two mutually perpendicular line elements in the tangent plane (the principal directions). The displacement as a rigid body is determined by the translation that carries P into the corresponding point P_1 in the deformed state, and by the rotation $\boldsymbol{\omega}$ associated with the contribution $\delta\mathbf{a}_\alpha^{(2)}$.

1.4.4 The bending of the middle surface

It was mentioned in section 1.2.2 that the geometry of the middle surface is determined by the first and second fundamental forms of the surface (except for a displacement as a rigid body). The change of shape of the middle surface as a result of the transition from the reference state to the deformed state (i.e. the deformation of the middle surface) is therefore determined by the changes of the coefficients of the two fundamental forms.

In section 1.4.2 we considered the first fundamental form, and we found that the quantities $(a^*_{\alpha\beta} - a_{\alpha\beta})$ describe the strains of the middle surface. We shall now determine the changes of the coefficients of the second fundamental form. From (1.18) we have

$$(b^*_{\alpha\beta} - b_{\alpha\beta})_{\text{lin}} = \delta b_{\alpha\beta} = -\delta(\mathbf{a}_{3,\alpha} \cdot \mathbf{a}_\beta)$$
$$= -\delta(\mathbf{a}_{3,\alpha}) \cdot \mathbf{a}_\beta - \mathbf{a}_{3,\alpha} \cdot \delta\mathbf{a}_\beta = -(\delta\mathbf{a}_3)_{,\alpha} \cdot \mathbf{a}_\beta - \mathbf{a}_{3,\alpha} \cdot \delta\mathbf{a}_\beta,$$

where we have used the rule (1.53) for the first variation. With the help of (1.74), (1.55) and (1.37) we further derive

$$\delta b_{\alpha\beta} = -(\boldsymbol{\omega} \times \mathbf{a}_3)_{,\alpha} \cdot \mathbf{a}_\beta + \frac{1}{R_\alpha}\mathbf{a}_\alpha \cdot \mathbf{v}_{,\beta}$$
$$= -\boldsymbol{\omega}_{,\alpha} \cdot \hat{\mathbf{a}}_\beta + \boldsymbol{\omega} \cdot \frac{1}{R_\alpha}\mathbf{a}_\alpha \times \mathbf{a}_\beta + \frac{1}{R_\alpha}\mathbf{a}_\alpha \cdot \mathbf{v}_{,\beta}$$
$$= -\boldsymbol{\omega}_{,\alpha} \cdot \hat{\mathbf{a}}_\beta + \frac{1}{R_\alpha}\mathbf{a}_\alpha \cdot (\mathbf{v}_{,\beta} - \boldsymbol{\omega} \times \mathbf{a}_\beta).$$

Transforming the last term by means of (1.76) we obtain

$$\delta b_{\alpha\beta} = -\boldsymbol{\omega}_{,\alpha} \cdot \hat{\mathbf{a}}_\beta + \frac{A_\alpha A_\beta}{R_\alpha}\varepsilon_{\alpha\beta}. \tag{1.84}$$

For the first term on the right-hand side we introduce the notation

$$A_\alpha A_\beta k_{\alpha\beta} = \boldsymbol{\omega}_{,\alpha} \cdot \hat{\mathbf{a}}_\beta, \tag{1.85}$$

so that

$$\delta b_{\alpha\beta} = A_\alpha A_\beta\left(-k_{\alpha\beta} + \frac{1}{R_\alpha}\varepsilon_{\alpha\beta}\right). \tag{1.86}$$

We notice that $\delta b_{\alpha\beta}$ is symmetric in the indices α and β (this follows from the fact that $(b^*_{\alpha\beta} - b_{\alpha\beta})$ is symmetric). We now consider the two equations obtained from (1.86) by putting (α, β) equal to (1, 2) and (2, 1), respectively. Subtracting one of these equations from the other we obtain

$$k_{12} - k_{21} = \left(\frac{1}{R_1} - \frac{1}{R_2}\right)\varepsilon_{12}. \tag{1.87}$$

We now form the quantity $\frac{1}{2}(\delta b_{\alpha\beta} + \delta b_{\beta\alpha})$ by means of (1.86) and deduce the following alternative expression for $\delta b_{\alpha\beta}$:

$$\delta b_{\alpha\beta} = A_\alpha A_\beta\left[-\kappa_{\alpha\beta} + \frac{1}{2}\left(\frac{1}{R_\alpha} + \frac{1}{R_\beta}\right)\varepsilon_{\alpha\beta}\right], \tag{1.88}$$

where we have introduced the quantities

$$\kappa_{\alpha\beta} = \tfrac{1}{2}(k_{\alpha\beta} + k_{\beta\alpha}) = \kappa_{\beta\alpha}, \tag{1.89}$$

which are symmetric in the indices α and β.

We notice that the quantities $\varepsilon_{\alpha\beta}$ and $\kappa_{\alpha\beta}$ have the following properties:

(1) It follows from (1.61) and (1.88) that, within the present linearized theory, the quantities $\varepsilon_{\alpha\beta}$ and $\kappa_{\alpha\beta}$ determine the changes of the coefficients of the first and second fundamental forms of the middle surface (and therefore, as previously explained, the geometry of the surface in the deformed state, apart from a displacement as a rigid body). In section 1.4.5 it will be shown directly that $\varepsilon_{\alpha\beta}$ and $\kappa_{\alpha\beta}$ determine the shape of the middle surface in the deformed state (provided that these quantities satisfy the so-called equations of compatibility (see section 1.4.5)).

(2) Let us consider an arbitrary infinitesimal displacement as a rigid body of the shell. Such a displacement can always be written in the form

$$\mathbf{v}(\theta_1, \theta_2) = \mathbf{v}_0 + \boldsymbol{\omega}_0 \times \mathbf{r}(\theta_1, \theta_2), \tag{1.90}$$

where $\mathbf{v}_0$ and $\boldsymbol{\omega}_0$ are constant vectors. Substituting in (1.55), we get

$$\delta\mathbf{a}_\alpha = \mathbf{v}_{,\alpha} = \boldsymbol{\omega}_0 \times \mathbf{r}_{,\alpha} = \boldsymbol{\omega}_0 \times \mathbf{a}_\alpha. \tag{1.91}$$

We notice that the rotation vector $\boldsymbol{\omega}(\theta_1, \theta_2)$ belonging to this displacement is given by

$$\boldsymbol{\omega} = \boldsymbol{\omega}_0 = \text{constant},$$

as this expression for $\boldsymbol{\omega}$, together with (1.91), satisfies the equations (1.81) and (1.82) which determine the vector $\boldsymbol{\omega}$ uniquely. We therefore have

$$\boldsymbol{\omega}_{,\alpha} = \boldsymbol{\omega}_{0,\alpha} = \mathbf{0}, \qquad \mathbf{v}_{,\alpha} - \boldsymbol{\omega} \times \mathbf{a}_\alpha = \boldsymbol{\omega}_0 \times \mathbf{a}_\alpha - \boldsymbol{\omega}_0 \times \mathbf{a}_\alpha = \mathbf{0},$$

and these equations together with (1.76) and (1.85) show that $\varepsilon_{\alpha\beta}$ and $\kappa_{\alpha\beta}$ vanish identically for any displacement as a rigid body.

The results (1) and (2) imply that *the six quantities $\varepsilon_{\alpha\beta}$ and $\kappa_{\alpha\beta}$ determine the change of shape and therefore the deformation of the shell.*

It was shown in section 1.4.2 that the quantities $\varepsilon_{\alpha\beta}$ determine the strains of the middle surface. In order to clarify the properties of the quantities $\kappa_{\alpha\beta}$, we shall now calculate the changes of the normal curvatures and the torsion of the surface along the coordinate curves, i.e. the linearized changes of the quantities (see also Fig. 9 and equation (1.14))

$$-\left(\frac{d\mathbf{a}_3}{ds_\alpha} \cdot \mathbf{e}_\beta\right) = -\left(\frac{\mathbf{a}_{3,\alpha}}{A_\alpha} \cdot \frac{\mathbf{a}_\beta}{A_\beta}\right). \tag{1.92}$$

These linearized changes are now calculated as the first variations, i.e.

$$-\delta\left(\frac{\mathbf{a}_{3,\alpha} \cdot \mathbf{a}_\beta}{A_\alpha A_\beta}\right) = \delta\left(\frac{b_{\alpha\beta}}{A_\alpha A_\beta}\right) = \frac{1}{A_\alpha A_\beta}\left[\delta b_{\alpha\beta} - b_{\alpha\beta}\left(\frac{\delta A_\alpha}{A_\alpha} + \frac{\delta A_\beta}{A_\beta}\right)\right].$$

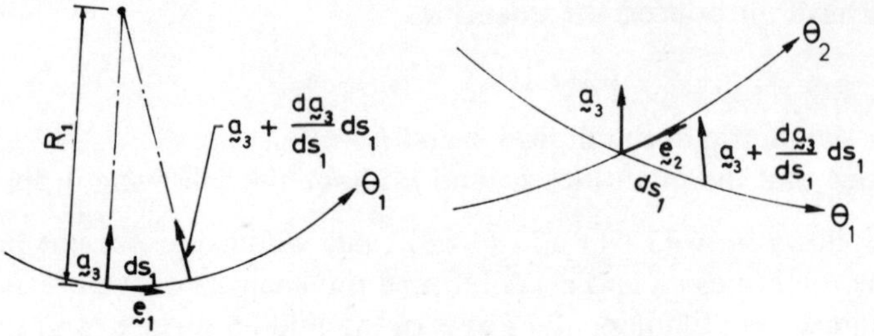

Fig. 9. Normal curvature and torsion

If we insert the expression (1.88) for $\delta b_{\alpha\beta}$ and use (1.67), we obtain

$$\delta\left(\frac{b_{\alpha\beta}}{A_\alpha A_\beta}\right) = -\kappa_{\alpha\beta} + \frac{1}{2}\left(\frac{1}{R_\alpha} + \frac{1}{R_\beta}\right)\varepsilon_{\alpha\beta} - \frac{b_{\alpha\beta}}{A_\alpha A_\beta}(\varepsilon_{\alpha\alpha} + \varepsilon_{\beta\beta}).$$

It follows from (1.41) that $b_{\alpha\beta}$ may be written in the form

$$b_{\alpha\beta} = A_\alpha A_\beta \frac{1}{2}\left(\frac{1}{R_\alpha} + \frac{1}{R_\beta}\right)\delta_{\alpha\beta},$$

where $\delta_{\alpha\beta}$ is the *Kronecker delta* defined by

$$\delta_{\alpha\beta} = \begin{cases} 1 & \text{for } \alpha = \beta, \\ 0 & \text{for } \alpha \neq \beta. \end{cases} \tag{1.93}$$

Substituting the last expression for $b_{\alpha\beta}$ in the equation that precedes it, we get

$$-\delta\left(\frac{\mathbf{a}_{3,\alpha} \cdot \mathbf{a}_\beta}{A_\alpha A_\beta}\right) = -\left[\kappa_{\alpha\beta} + \frac{1}{2}\left(\frac{1}{R_\alpha} + \frac{1}{R_\beta}\right)[(\varepsilon_{\alpha\alpha} + \varepsilon_{\beta\beta})\delta_{\alpha\beta} - \varepsilon_{\alpha\beta}]\right], \tag{1.94}$$

or in full:

Changes of normal curvatures

$$\delta(b_{11}/A_1^2) = -(\kappa_{11} + \varepsilon_{11}/R_1),$$
$$\delta(b_{22}/A_2^2) = -(\kappa_{22} + \varepsilon_{22}/R_2),$$

Changes of torsion

$$\delta\left(\frac{b_{12}}{A_1 A_2}\right) = -\left[\kappa_{12} - \tfrac{1}{2}\varepsilon_{12}\left(\frac{1}{R_1} + \frac{1}{R_2}\right)\right].$$

It will be seen that these quantities (the *changes of curvature* of the surface) are determined by $\varepsilon_{\alpha\beta}$ and $\kappa_{\alpha\beta}$. It follows from (1.85) and (1.89) that $\kappa_{\alpha\beta}$ is completely determined by the derivatives $\boldsymbol{\omega}_{,\alpha}$ of the rotation vector. The results of our considerations may therefore be summarized in the following manner: The changes of curvature of the surface can be written as a sum of two contributions. One of these (namely $\kappa_{\alpha\beta}$) depends only on the rotation vector, whereas the second contribution depends on the strain measures $\varepsilon_{\alpha\beta}$

and vanishes in the case of inextensional displacements. The quantities $\kappa_{\alpha\beta}$ are called the *bending measures* of the middle surface.

Several different versions of the theory of shells have been proposed in the literature. While the quantities $\varepsilon_{\alpha\beta}$ are universally accepted as strain measures, the theories differ with respect to the quantities used as bending measures. For example, the changes of the coefficients of the second fundamental form $\delta b_{\alpha\beta}$ may be used as bending measures, since it follows from previous remarks that the quantities $\varepsilon_{\alpha\beta}$ and $\delta b_{\alpha\beta}$ determine the change of shape of the shell. With the help of these strain and bending measures, a consistent theory of shells may then be developed. Our choice of the quantities $\kappa_{\alpha\beta}$ as bending measures in the present derivation of the theory is motivated by the fact that the resulting theory possesses a number of advantageous properties, as will be shown more fully in section 1.8.

From (1.85) and (1.89) we derive the following expression for the bending measures:

$$\kappa_{\alpha\beta} = \frac{1}{2A_\alpha A_\beta}(\boldsymbol{\omega}_{,\alpha}\cdot\hat{\mathbf{a}}_\beta + \boldsymbol{\omega}_{,\beta}\cdot\hat{\mathbf{a}}_\alpha). \tag{1.95}$$

Using (1.40) we now express $\kappa_{\alpha\beta}$ in terms of the components ω_i of the rotation vector (note that $\boldsymbol{\omega} = -\omega_2\mathbf{e}_1 + \omega_1\mathbf{e}_2 + \omega_3\mathbf{e}_3$; see (1.80)), and we then insert the expressions (1.83) for ω_i so as to express the bending measures in terms of the displacements. Thus

$$\begin{aligned}
\kappa_{11} &= \frac{1}{A_1}\left(\omega_{1,1} + \frac{A_{1,2}}{A_2}\omega_2\right) = -\frac{1}{A_1}\left(\frac{v_{3,1}}{A_1} + \frac{v_1}{R_1}\right)_{,1} - \frac{A_{1,2}}{A_1A_2}\left(\frac{v_{3,2}}{A_2} + \frac{v_2}{R_2}\right),\\
\kappa_{22} &= \frac{1}{A_2}\left(\omega_{2,2} + \frac{A_{2,1}}{A_1}\omega_1\right) = -\frac{1}{A_2}\left(\frac{v_{3,2}}{A_2} + \frac{v_2}{R_2}\right)_{,2} - \frac{A_{2,1}}{A_1A_2}\left(\frac{v_{3,1}}{A_1} + \frac{v_1}{R_1}\right),\\
\kappa_{12} &= \frac{1}{2}\left[\frac{A_2}{A_1}\left(\frac{\omega_2}{A_2}\right)_{,1} + \frac{A_1}{A_2}\left(\frac{\omega_1}{A_1}\right)_{,2} + \left(\frac{1}{R_1} - \frac{1}{R_2}\right)\omega_3\right]\\
&= -\frac{1}{A_1A_2}\left(v_{3,12} - \frac{A_{1,2}}{A_1}v_{3,1} - \frac{A_{2,1}}{A_2}v_{3,2}\right) - \frac{1}{4}\left(\frac{3}{R_1} - \frac{1}{R_2}\right)\frac{A_1}{A_2}\left(\frac{v_1}{A_1}\right)_{,2}\\
&\quad - \frac{1}{4}\left(\frac{3}{R_2} - \frac{1}{R_1}\right)\frac{A_2}{A_1}\left(\frac{v_2}{A_2}\right)_{,1},
\end{aligned} \tag{1.96}$$

where we have used Codazzi's equations (1.44) in the derivation of the final expression for κ_{12}. Note that in the limiting case of a plane plate and rectangular Cartesian coordinates in the plane we have $A_1 = A_2 = 1$, and $1/R_1 = 1/R_2 = 0$. The expressions for the bending measures then reduce to $\kappa_{\alpha\beta} = -v_{3,\alpha\beta}$, which is the usual formula for the changes of curvature in the theory of bending of plane plates.

1.4.5 Equations of compatibility

The equations (1.66) and (1.96) determine the strain and bending measures as functions of the displacements. If the displacements are given, these equations can therefore be used to calculate the corresponding strain and bending measures. We shall now consider the converse problem, namely the determination of the displacements belonging to given strain and bending measures, the reference state being known. We therefore assume that $\varepsilon_{\alpha\beta}$ and $\kappa_{\alpha\beta}$ are given functions satisfying the symmetry conditions (1.63) and (1.89), and we wish to determine the corresponding displacement $\mathbf{v}$. This will be done in two stages. The first of these involves determination of the rotation vector $\boldsymbol{\omega}$, while the second stage yields the required expression for $\mathbf{v}$.

We notice that the quantities $k_{\alpha\beta}$ are determined by $\varepsilon_{\alpha\beta}$ and $\kappa_{\alpha\beta}$. This follows from the equation

$$k_{\alpha\beta} = \kappa_{\alpha\beta} + \frac{1}{2}\left(\frac{1}{R_\alpha} - \frac{1}{R_\beta}\right)\varepsilon_{\alpha\beta}, \tag{1.97}$$

which can be verified by comparison with (1.89) and (1.87). We also note that the vectors $\boldsymbol{\omega}_{,\alpha}$ and $\mathbf{v}_{,\alpha}$ can be represented in the following manner:

$$\begin{aligned} \boldsymbol{\omega}_{,\alpha} &= A_\alpha \sum_\beta k_{\alpha\beta}\hat{\mathbf{e}}_\beta + k_\alpha \mathbf{e}_3, \\ \mathbf{v}_{,\alpha} &= \boldsymbol{\omega} \times \mathbf{a}_\alpha + A_\alpha \sum_\beta \varepsilon_{\alpha\beta}\mathbf{e}_\beta. \end{aligned} \tag{1.98}$$

As far as the first of these equations is concerned, the components of the vector $\boldsymbol{\omega}_{,\alpha}$ in the direction of the unit vectors $\hat{\mathbf{e}}_\beta$ are given by (1.85), and the component in the direction of $\mathbf{e}_3$ is denoted by k_α. The second equation follows from (1.75). It will be convenient to introduce the notations

$$\mathbf{k}_\alpha = A_\alpha \sum_\beta k_{\alpha\beta}\hat{\mathbf{e}}_\beta + k_\alpha \mathbf{e}_3, \qquad \boldsymbol{\gamma}_\alpha = A_\alpha \sum_\beta \varepsilon_{\alpha\beta}\mathbf{e}_\beta \tag{1.99}$$

for the right-hand side of $(1.98)_1$ and for the last term on the right-hand side of $(1.98)_2$. We then have the equations

$$\boldsymbol{\omega}_{,\alpha} = \mathbf{k}_\alpha, \qquad \mathbf{v}_{,\alpha} = \boldsymbol{\omega} \times \mathbf{a}_\alpha + \boldsymbol{\gamma}_\alpha \tag{1.100}$$

for determination of $\boldsymbol{\omega}$ and $\mathbf{v}$. Now it follows from (1.99) that when $\varepsilon_{\alpha\beta}$ and $\kappa_{\alpha\beta}$ (and therefore also $k_{\alpha\beta}$) are known functions, the quantities $\boldsymbol{\gamma}_\alpha$ are also known, but the quantities $\mathbf{k}_\alpha$ are only partially known, because the components k_α have not yet been determined. However, it will now be shown that k_α can be expressed in terms of the strain measures $\varepsilon_{\alpha\beta}$.

Let us for a moment suppose that the $\mathbf{k}_\alpha$ are known functions of (θ_1, θ_2). We would then seek to determine $\boldsymbol{\omega}$ by integration of $(1.100)_1$, since the right-hand side of this equation would be a known function. Substituting the resulting expression for $\boldsymbol{\omega}$ in the right-hand side of $(1.100)_2$, we would determine $\mathbf{v}$ by integration of this equation. However, in order for this procedure to be feasible, it is well known that the following conditions of integrability must be satisfied:

$$\mathbf{k}_{1,2} = \mathbf{k}_{2,1}, \qquad (\boldsymbol{\omega} \times \mathbf{a}_1 + \boldsymbol{\gamma}_1)_{,2} = (\boldsymbol{\omega} \times \mathbf{a}_2 + \boldsymbol{\gamma}_2)_{,1}. \tag{1.101}$$

These equations constitute the necessary and sufficient conditions for the existence of single-valued functions $\boldsymbol{\omega}$ and $\mathbf{v}$, the partial derivatives of which are given by the expressions on the right-hand sides of (1.100) (assuming that the region Ω in the (θ_1, θ_2)-plane is simply connected).

We now perform the differentiation in $(1.101)_2$, replace $\boldsymbol{\omega}_{,\alpha}$ by $\mathbf{k}_\alpha$ in the resulting expression (see $(1.100)_1$), and use the fact that $\mathbf{a}_{1,2} = \mathbf{a}_{2,1}$ (see (1.5)). Equation $(1.101)_2$ then assumes the form

$$\mathbf{k}_1 \times \mathbf{a}_2 - \mathbf{k}_2 \times \mathbf{a}_1 = \gamma_{1,2} - \gamma_{2,1}. \tag{1.101a}$$

Substituting $(1.99)_2$ in the condition (1.101a), we then obtain

$$\mathbf{k}_1 \times \mathbf{a}_2 - \mathbf{k}_2 \times \mathbf{a}_1 = [A_1(\varepsilon_{11}\mathbf{e}_1 + \varepsilon_{12}\mathbf{e}_2)]_{,2} - [A_2(\varepsilon_{21}\mathbf{e}_1 + \varepsilon_{22}\mathbf{e}_2)]_{,1}. \tag{1.102}$$

We now insert the expression $(1.99)_1$ for $\mathbf{k}_\alpha$ in the left-hand side of (1.102) and calculate the derivatives on the right-hand side of the latter equation by means of (1.40). The resulting vector equation corresponds to three scalar equations. One of these is identical to (1.87), while the other two equations assume the form

$$\begin{aligned} k_1 &= -\frac{1}{A_2}[(A_1\varepsilon_{11})_{,2} - (A_2\varepsilon_{12})_{,1} - A_{2,1}\varepsilon_{12} - A_{1,2}\varepsilon_{22}], \\ k_2 &= \frac{1}{A_1}[(A_2\varepsilon_{22})_{,1} - (A_1\varepsilon_{12})_{,2} - A_{1,2}\varepsilon_{12} - A_{2,1}\varepsilon_{11}]. \end{aligned} \tag{1.103}$$

We see that these equations *express the quantities k_α in terms of $\varepsilon_{\alpha\beta}$*. Substituting the expression $(1.99)_1$ for $\mathbf{k}_\alpha$ in $(1.101)_1$ we obtain

$$-A_1k_{12}\mathbf{e}_1 + A_1k_{11}\mathbf{e}_2 + k_1\mathbf{e}_3]_{,2} = [-A_2k_{22}\mathbf{e}_1 + A_2k_{21}\mathbf{e}_2 + k_2\mathbf{e}_3]_{,1}.$$

The derivatives in this equation are now evaluated by means of (1.40), and we also introduce the expressions (1.103) for k_α and (1.97) for $k_{\alpha\beta}$. In this way the following three equations are derived:

$$\begin{aligned} &-(A_2\kappa_{22})_{,1} + (A_1\kappa_{12})_{,2} + A_{1,2}\kappa_{12} + A_{2,1}\kappa_{11} - (1/R_1)[(A_2\varepsilon_{22})_{,1} \\ &-(A_1\varepsilon_{12})_{,2} - A_{1,2}\varepsilon_{12} - A_{2,1}\varepsilon_{11}] + \tfrac{1}{2}A_1[\varepsilon_{12}(1/R_1 - 1/R_2)]_{,2} = 0, \\ &-(A_1\kappa_{11})_{,2} + (A_2\kappa_{12})_{,1} + A_{2,1}\kappa_{12} + A_{1,2}\kappa_{22} - (1/R_2)[(A_1\varepsilon_{11})_{,2} \\ &-(A_2\varepsilon_{12})_{,1} - A_{2,1}\varepsilon_{12} - A_{1,2}\varepsilon_{22}] + \tfrac{1}{2}A_2[\varepsilon_{12}(1/R_2 - 1/R_1)]_{,1} = 0, \\ &\{(1/A_1)[(A_2\varepsilon_{22})_{,1} - (A_1\varepsilon_{12})_{,2} - A_{1,2}\varepsilon_{12} - A_{2,1}\varepsilon_{11}]\}_{,1} \\ &+ \{(1/A_2)[(A_1\varepsilon_{11})_{,2} - (A_2\varepsilon_{12})_{,1} - A_{2,1}\varepsilon_{12} - A_{1,2}\varepsilon_{22}]\}_{,2} \\ &-A_1A_2(\kappa_{11}/R_2 + \kappa_{22}/R_1) = 0. \end{aligned} \tag{1.104}$$

These three equations are called the *equations of compatibility* of the shell. It can be shown that the equations of compatibility are, in fact, identical to the equations of Gauss and Codazzi for the middle surface in the deformed state.

It follows from the derivation that the conditions (1.101) imply the validity of the equations (1.87), (1.103) and (1.104). It will be seen that the equations (1.104) represent conditions that must be satisfied by the strain and bending measures, whereas equations (1.87) and (1.103) do not impose any restrictions on these quantities and may be regarded as definitions of $(k_{12} - k_{21})$ and k_α in terms of $\varepsilon_{\alpha\beta}$ and $\kappa_{\alpha\beta}$. It follows that if $\varepsilon_{\alpha\beta}$ and $\kappa_{\alpha\beta}$ satisfy the equations of compatibility (1.104), the conditions (1.101) will also be satisfied, and $\boldsymbol{\omega}$ and $\mathbf{v}$ can then be determined by integration of (1.100). On carrying out the integration, we find that the resulting expression for $\mathbf{v}$ contains a term of the type $(\mathbf{v}_0 + \boldsymbol{\omega}_0 \times \mathbf{r})$, where $\mathbf{v}_0$ and $\boldsymbol{\omega}_0$ are arbitrary constant vectors. It follows that the strain and bending measures determine the displacements except for an arbitrary displacement as a rigid body.

1.5 Statics of the shell

In the previous sections we have dealt with the geometrical description of the shell (displacements and deformations). In the present section we shall consider the equilibrium conditions of the shell.

In the theory of shells we are usually concerned with the following problem. We wish to determine the equilibrium state of the shell under the action of a given applied load. The reference state is known and the deformations and displacements of the shell depend on the external load. The configuration of the shell in the equilibrium state (the deformed state) is therefore initially unknown, and determination of this configuration is an important part of the solution of the above problem.

It was assumed in the geometrical description that the displacements of the shell were infinitesimal. This assumption will be retained in the derivation of the equilibrium conditions. It follows that the configurations of the shell in the reference state and in the deformed state only differ by infinitesimal quantities. We shall now introduce the further assumption that it is permissible to neglect the difference between the two configurations for the purpose of deriving the equilibrium conditions. This means that we shall *use the known geometry of the reference state when formulating the equilibrium conditions*. This approximation considerably simplifies the theory.

While it is true that several types of problems in the theory of shells can be treated in this way with sufficient accuracy, it should be emphasized that there are cases (e.g. problems in the theory of stability of shells) in which it is essential that the change of geometry be taken into account when the equilibrium conditions are formulated.

1.5.1 Applied loads, contact forces and couples

It is assumed that the shell is in equilibrium under the action of the following forces and moments:

(a) *Applied loads* consisting of a distributed surface load given by the force intensity **p** per unit area of the middle surface, and of loads acting along the edge of the shell. The edge loads will be dealt with in more detail in section 1.7.

(b) *Contact forces and couples*. Consider a point P of the middle surface and a curve on the surface through this point. We introduce a unit normal vector **ν** to the curve, i.e. a unit vector in the tangent plane at P which is perpendicular to the tangent to the curve at this point (see Fig. 10).

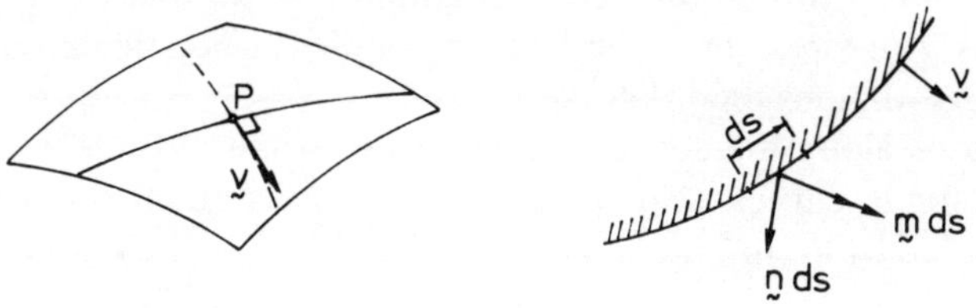

Fig. 10. Normal vector to curve, and contact force and couple

We now consider a section through the two-dimensional body (the shell) along the curve, and that part of the shell for which **ν** is the exterior normal (i.e. the shaded part in Fig. 10). Consider also an infinitesimal element of the section curve of length d*s* at the point P. It is now assumed that an internal force **n** d*s* and an internal moment **m** d*s* act on this element and on the shaded part of the shell (the force and moment on the corresponding element of the unshaded part are then given by −**n** d*s* and −**m** d*s*, respectively). The vectors **n** and **m** are called the *contact force* vector and the *contact couple* vector, respectively, and they are evidently force and moment intensities measured per unit length of section curve. We shall generally use the terms *contact forces and couples* to denote *internal force and moment intensities*. It is assumed that **n** and **m** are the same for all section curves through P having the same tangent at this point.

It is also assumed that *the contact couple vector* **m** *at* P *lies in the corresponding tangent plane* to the middle surface. In order to motivate this assumption we shall, for a moment, regard the shell as a three-dimensional body, for which the section curve is replaced by a section surface composed of the normals to the middle surface along the curve. We consider a small element of this surface corresponding to the line element d*s* of the curve (see Fig. 11).

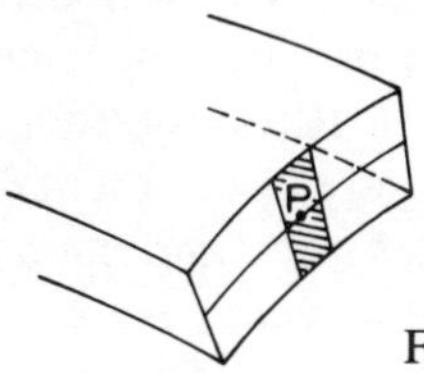

Fig. 11. Section through three-dimensional shell

The internal force $\mathbf{n}\,ds$ and moment $\mathbf{m}\,ds$ may now be interpreted as the resultant force and the resultant moment about P due to the stresses on the surface element. In the same way as in the theory of bending of plane plates, it can now be seen that the moment about the surface normal at P is a small quantity of second order in ds. The corresponding moment component per unit length of section curve (which is obtained by dividing the moment by ds and then letting $ds \to 0$) therefore vanishes.

In the present two-dimensional version of the theory of shells, we shall use the terms 'contact forces' and 'contact couples' in preference to the conventional terms 'stress resultants' and 'stress couples', since the latter suggest the use of a three-dimensional theory.

For sections along the coordinate curves we shall put $\boldsymbol{\nu} = \mathbf{e}_1$ for θ_2-curves and $\boldsymbol{\nu} = \mathbf{e}_2$ for θ_1-curves, and the corresponding contact forces and couples will be denoted by $\mathbf{n}_1$, $\mathbf{m}_1$, $\mathbf{n}_2$, and $\mathbf{m}_2$ (see Fig. 12). For these sections it will be

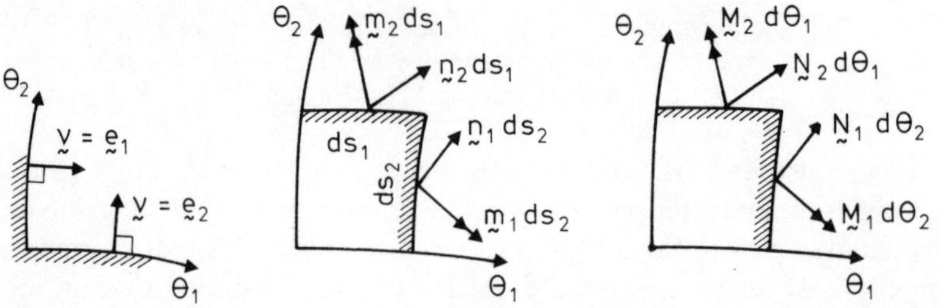

Fig. 12. Contact forces and couples acting on sections along coordinate curves

convenient to introduce the quantities $\mathbf{N}_\alpha$ and $\mathbf{M}_\alpha$ which are proportional to $\mathbf{n}_\alpha$ and $\mathbf{m}_\alpha$ and which are defined by

$$\mathbf{n}_1\, ds_2 = \mathbf{N}_1\, d\theta_2, \qquad \mathbf{m}_1\, ds_2 = \mathbf{M}_1\, d\theta_2,$$

etc., or, by using the relations $ds_\alpha = A_\alpha\, d\theta_\alpha$ and $G = A_1 A_2$,

$$\mathbf{n}_1 = \frac{\mathbf{N}_1}{A_2} = \frac{A_1}{G}\mathbf{N}_1, \qquad \mathbf{m}_1 = \frac{\mathbf{M}_1}{A_2} = \frac{A_1}{G}\mathbf{M}_1,$$

etc., so that we have

$$\mathbf{n}_\alpha = \frac{A_\alpha}{G}\mathbf{N}_\alpha, \qquad \mathbf{m}_\alpha = \frac{A_\alpha}{G}\mathbf{M}_\alpha. \tag{1.105}$$

It will be seen that the internal force and moment acting on a small element of a section along a θ_2-curve corresponding to the increment $d\theta_2$ are given by $\mathbf{N}_1\, d\theta_2$ and $\mathbf{M}_1\, d\theta_2$, respectively (see Fig. 12). It follows that $\mathbf{N}_1$ and $\mathbf{M}_1$ are intensities of internal force and moment measured per unit of θ_2, and similarly for $\mathbf{N}_2$ and $\mathbf{M}_2$, so that $\mathbf{N}_\alpha$ and $\mathbf{M}_\alpha$ are contact force and couple vectors.

Variation of contact forces and couples at a point. We consider again the small line element ds of the section curve through P. Let the endpoints of the element be denoted by P′ and Q′, and let the direction P′Q′ be the same as the direction determined by the unit tangent vector

$$\mathbf{t} = \mathbf{a}_3 \times \boldsymbol{\nu} \tag{1.106}$$

to the section curve at P (see Fig. 13). We now consider a small triangular area element of the middle surface bounded by the section P′Q′ and by sections along coordinate curves through P′ and Q′ so that $\boldsymbol{\nu}$ is the outer normal. Let the curvilinear coordinates of the points P′ and Q′ be θ_α and $(\theta_\alpha + \mathrm{d}\theta_\alpha)$, respectively. Fig. 13 shows the internal forces and moments which act on the element.

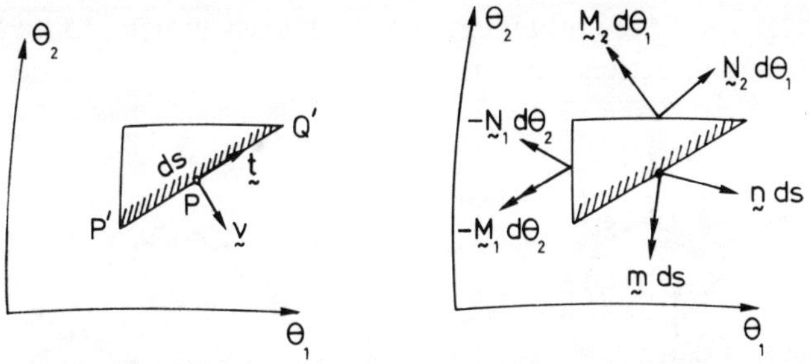

Fig. 13. Variation of contact forces and couples at a point

We now formulate the equilibrium conditions for the element (force equilibrium and moment equilibrium about point P), divide the resulting expressions by ds, and let d$s \to 0$. In this way we obtain the formulae

$$\mathbf{n} = \mathbf{N}_1 \frac{\mathrm{d}\theta_2}{\mathrm{d}s} - \mathbf{N}_2 \frac{\mathrm{d}\theta_1}{\mathrm{d}s}, \qquad \mathbf{m} = \mathbf{M}_1 \frac{\mathrm{d}\theta_2}{\mathrm{d}s} - \mathbf{M}_2 \frac{\mathrm{d}\theta_1}{\mathrm{d}s}, \tag{1.107}$$

since the contributions from the distributed load $\mathbf{p}$ and the contribution from $\mathbf{N}_\alpha$ to the moment equation are infinitesimals of a higher order which tend to zero as d$s \to 0$. The equations (1.107) determine the contact force and couple on an arbitrary section through P as functions of the contact forces and couples belonging to the coordinate curves.

The components of the contact forces and couples. We now resolve the contact force $\mathbf{n}_1$ in components in the directions of the unit vectors $\mathbf{e}_i$. Thus

$$\mathbf{n}_1 = \frac{A_1}{G} \mathbf{N}_1 = \mathrm{N}_{11}\mathbf{e}_1 + \mathrm{N}_{12}\mathbf{e}_2 + \mathrm{Q}_1\mathbf{e}_3.$$

The other contact forces and couples belonging to the coordinate curves are treated in a similar manner, so that we obtain the formulae

$$\begin{aligned} \mathbf{N}_1 &= \frac{G}{A_1}(\mathrm{N}_{11}\mathbf{e}_1 + \mathrm{N}_{12}\mathbf{e}_2 + \mathrm{Q}_1\mathbf{e}_3), \qquad & \mathbf{M}_1 &= \frac{G}{A_1}(\mathrm{M}_{11}\hat{\mathbf{e}}_1 + \mathrm{M}_{12}\hat{\mathbf{e}}_2), \\ \mathbf{N}_2 &= \frac{G}{A_2}(\mathrm{N}_{21}\mathbf{e}_1 + \mathrm{N}_{22}\mathbf{e}_2 + \mathrm{Q}_2\mathbf{e}_3), \qquad & \mathbf{M}_2 &= \frac{G}{A_2}(\mathrm{M}_{21}\hat{\mathbf{e}}_1 + \mathrm{M}_{22}\hat{\mathbf{e}}_2), \end{aligned} \tag{1.108}$$

or briefly

$$\mathbf{N}_\alpha = \frac{G}{A_\alpha}\left(\sum_\beta \mathrm{N}_{\alpha\beta}\mathbf{e}_\beta + \mathrm{Q}_\alpha\mathbf{e}_3\right),$$
$$\mathbf{M}_\alpha = \frac{G}{A_\alpha}\sum_\beta \mathrm{M}_{\alpha\beta}\hat{\mathbf{e}}_\beta, \tag{1.109}$$

in which *the components denote forces and moments per unit length of coordinate curve*. Note that the moment vectors are resolved in the directions of the transverse vectors $\hat{\mathbf{e}}_\alpha$ (see (1.79a)). It follows from the formulae (1.108) that positive values of the components imply that the contact forces and couples on a small element of the middle surface with sides parallel to the coordinate curves act in the directions shown in Fig. 14.

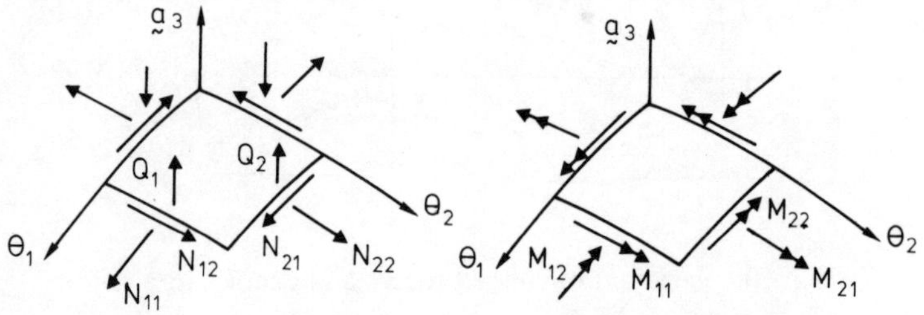

Fig. 14. Components of contact forces and couples

1.5.2 Equations of equilibrium

We consider an infinitesimal element of the middle surface bounded by coordinate curves. Fig. 15 shows the element and the relevant forces and moments which act on the element. The contribution from the applied load is obtained as the product of the force intensity **p** and the area $G\ d\theta_1\ d\theta_2$ of the element.

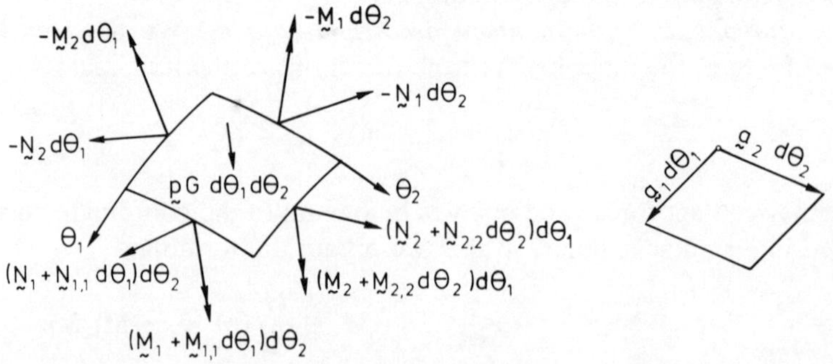

Fig. 15. Forces and couples acting on element of shell

We now formulate the equilibrium conditions for the element (force equilibrium and moment equilibrium about point P). Dividing the resulting equations by $d\theta_1\, d\theta_2$ we obtain the following *vector equations of equilibrium*:

Projection

$$\sum_{\alpha} \mathbf{N}_{\alpha,\alpha} + G\mathbf{p} = \mathbf{0},$$

Moments (1.110)

$$\sum_{\alpha} (\mathbf{M}_{\alpha,\alpha} + \mathbf{a}_\alpha \times \mathbf{N}_\alpha) = \mathbf{0}.$$

We shall now derive six scalar equations corresponding to the two vector equations (1.110). Inserting the expressions (1.108) for $\mathbf{N}_\alpha$ and $\mathbf{M}_\alpha$ in (1.110) and resolving $\mathbf{p}$ in the directions of the unit vectors $\mathbf{e}_i$ we get

$$[A_2(\mathrm{N}_{11}\mathbf{e}_1 + \mathrm{N}_{12}\mathbf{e}_2 + \mathrm{Q}_1\mathbf{e}_3)]_{,1} + [A_1(\mathrm{N}_{21}\mathbf{e}_1 + \mathrm{N}_{22}\mathbf{e}_2 + \mathrm{Q}_2\mathbf{e}_3)]_{,2} + G\sum_i p_i\mathbf{e}_i = \mathbf{0},$$

$$[A_2(-\mathrm{M}_{12}\mathbf{e}_1 + \mathrm{M}_{11}\mathbf{e}_2)]_{,1} + [A_1(-\mathrm{M}_{22}\mathbf{e}_1 + \mathrm{M}_{21}\mathbf{e}_2)]_{,2} + G(\mathrm{N}_{12} - \mathrm{N}_{21})\mathbf{e}_3 + G(\mathrm{Q}_2\mathbf{e}_1 - \mathrm{Q}_1\mathbf{e}_2) = \mathbf{0},$$

where we have used (1.79a) and the fact that $\mathbf{e}_1 \times \mathbf{e}_1 = \mathbf{0}$, $\mathbf{e}_1 \times \mathbf{e}_2 = \mathbf{e}_3$, etc. Evaluating the derivatives of the vectors in the brackets by means of the formulae (1.40), we obtain the desired six *scalar equations of equilibrium*:

Force equilibrium

$$\frac{1}{A_1A_2}[(A_2\mathrm{N}_{11})_{,1} + (A_1\mathrm{N}_{21})_{,2} + A_{1,2}\mathrm{N}_{12} - A_{2,1}\mathrm{N}_{22}] - \frac{\mathrm{Q}_1}{R_1} + p_1 = 0,$$

$$\frac{1}{A_1A_2}[(A_1\mathrm{N}_{22})_{,2} + (A_2\mathrm{N}_{12})_{,1} + A_{2,1}\mathrm{N}_{21} - A_{1,2}\mathrm{N}_{11}] - \frac{\mathrm{Q}_2}{R_2} + p_2 = 0, \qquad (1.111a)$$

$$\frac{\mathrm{N}_{11}}{R_1} + \frac{\mathrm{N}_{22}}{R_2} + \frac{1}{A_1A_2}[(A_2\mathrm{Q}_1)_{,1} + (A_1\mathrm{Q}_2)_{,2}] + p_3 = 0.$$

Moment equilibrium

$$\frac{1}{A_1A_2}[(A_2\mathrm{M}_{11})_{,1} + (A_1\mathrm{M}_{21})_{,2} + A_{1,2}\mathrm{M}_{12} - A_{2,1}\mathrm{M}_{22}] - \mathrm{Q}_1 = 0,$$

$$\frac{1}{A_1A_2}[(A_1\mathrm{M}_{22})_{,2} + (A_2\mathrm{M}_{12})_{,1} + A_{2,1}\mathrm{M}_{21} - A_{1,2}\mathrm{M}_{11}] - \mathrm{Q}_2 = 0, \qquad (1.111b)$$

$$\mathrm{N}_{12} - \mathrm{N}_{21} - \frac{\mathrm{M}_{12}}{R_1} + \frac{\mathrm{M}_{21}}{R_2} = 0.$$

It follows from the last equation that N_{12} is generally different from N_{21}. Likewise, there is no *a priori* reason for assuming that the moments $\mathrm{M}_{\alpha\beta}$ are symmetrical in the indices α and β.

1.5.3 The principle of virtual work

We assume that the region of the middle surface occupied by the two-dimensional body (the shell) is simply connected and bounded by a closed curve C consisting of a finite number of smooth arcs. The curve C is called the *boundary curve* of the shell. If $\boldsymbol{\nu}$ is the outer normal to the curve, the orientation of the curve is chosen in accordance with the direction determined by the tangent vectors **t** (see (1.106) and Fig. 16).

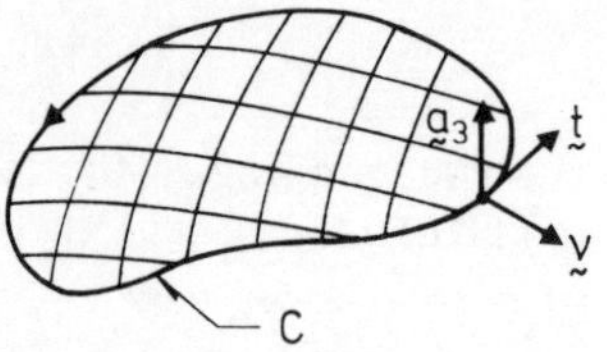

Fig. 16. Boundary curve of shell

We assume that the shell is in equilibrium under the action of the applied distributed load **p**, applied edge loads **n** and **m** (force and moment per unit length of boundary curve), and corresponding contact forces and couples $\mathbf{N}_\alpha$ and $\mathbf{M}_\alpha$. We also assume that the relations between the contact forces and couples along C and the applied edge loads **n**, **m** are given by the equilibrium conditions (1.107) (this implies that **m** lies in the tangent plane). We shall call these values of **n**, **m** the *equilibrium values* of the edge loads corresponding to the above contact forces and couples.

We now consider an arbitrary or *virtual infinitesimal displacement* **v** of the shell and the corresponding rotation vector $\boldsymbol{\omega}$ defined by (1.81) and (1.82). We assume that the applied, distributed loads and the edge loads act on the shell in such a manner that the work done by these loads in the infinitesimal displacement (the *external work*) is given by

$$A_{(e)} = \int \mathbf{p} \cdot \mathbf{v} \, dA + \oint_C (\mathbf{n} \cdot \mathbf{v} + \mathbf{m} \cdot \boldsymbol{\omega}) \, ds, \tag{1.112}$$

where the first integral is a surface integral taken over the part of the middle surface bounded by C, and the second integral is a line integral along the boundary curve C. The terms $\mathbf{p} \cdot \mathbf{v} \, dA$ and $\mathbf{n} \cdot \mathbf{v} \, ds$ in the integrands have the form that one would expect (the scalar product of a force vector with the corresponding displacement vector). However, some comments are called for in connection with the work of the applied moments,

$$\oint_C \mathbf{m} \cdot \boldsymbol{\omega} \, ds,$$

along the boundary curve.

Now when the theory of shells is derived with the help of the theory of three-dimensional bodies, the so-called *Love–Kirchhoff* hypothesis is often invoked. The most important part of this hypothesis is the assumption that straight fibres of the shell that are perpendicular to the middle surface before deformation remain so after deformation and do not change their length. It can be shown (see for example Goldenveizer[6]) that this hypothesis implies that the work of the applied moments along the boundary curve is given exactly by the expression

$$\oint_C \mathbf{m} \cdot \boldsymbol{\omega} \, \mathrm{d}s,$$

which appears in (1.112). The use of this expression for the work of the applied moments in conjunction with the definitions (1.81) and (1.82) of the components of the rotation vector may therefore be said to correspond to the introduction of the Love–Kirchhoff hypothesis.

We shall now derive the principle of virtual work by a transformation of the right-hand side of (1.112). For this purpose, we shall use *Green's theorem* in the form

$$\iint_\Omega (F_{1,1} + F_{2,2}) \, \mathrm{d}\theta_1 \, \mathrm{d}\theta_2 = \oint_c (F_1 \, \mathrm{d}\theta_2 - F_2 \, \mathrm{d}\theta_1), \qquad (1.113)$$

where F_1 and F_2 are functions of (θ_1, θ_2) defined in the region Ω, which is bounded by the curve c (the curve in the (θ_1, θ_2)-plane corresponding to the boundary curve C; see Fig. 17). We now substitute the expressions (1.107) for **n** and **m** in the right-hand side of (1.112) and use Green's theorem. Thus

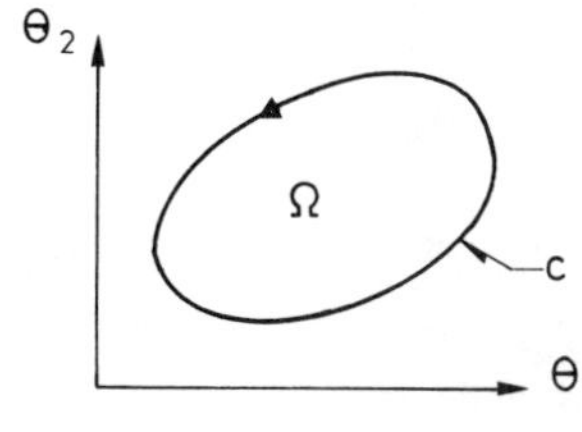

Fig. 17. Region Ω in (θ_1, θ_2)-plane

$$A_{(e)} = \int \mathbf{p} \cdot \mathbf{v} \, \mathrm{d}A + \oint_C \left[\left(\mathbf{N}_1 \frac{\mathrm{d}\theta_2}{\mathrm{d}s} - \mathbf{N}_2 \frac{\mathrm{d}\theta_1}{\mathrm{d}s} \right) \cdot \mathbf{v} + \left(\mathbf{M}_1 \frac{\mathrm{d}\theta_2}{\mathrm{d}s} - \mathbf{M}_2 \frac{\mathrm{d}\theta_1}{\mathrm{d}s} \right) \cdot \boldsymbol{\omega} \right] \mathrm{d}s$$

$$= \int \mathbf{p} \cdot \mathbf{v} \, \mathrm{d}A + \oint_c (\mathbf{N}_1 \cdot \mathbf{v} \, \mathrm{d}\theta_2 - \mathbf{N}_2 \cdot \mathbf{v} \, \mathrm{d}\theta_1 + \mathbf{M}_1 \cdot \boldsymbol{\omega} \, \mathrm{d}\theta_2 - \mathbf{M}_2 \cdot \boldsymbol{\omega} \, \mathrm{d}\theta_1)$$

$$= \int \mathbf{p} \cdot \mathbf{v} \, \mathrm{d}A + \iint_\Omega \left(\sum_\alpha (\mathbf{N}_\alpha \cdot \mathbf{v})_{,\alpha} + \sum_\alpha (\mathbf{M}_\alpha \cdot \boldsymbol{\omega})_{,\alpha} \right) \mathrm{d}\theta_1 \, \mathrm{d}\theta_2$$

$$= \sum_\alpha \int [\mathbf{N}_\alpha \cdot \mathbf{v}_{,\alpha} + \mathbf{M}_\alpha \cdot \boldsymbol{\omega}_{,\alpha} + (\mathbf{N}_{\alpha,\alpha} + G\mathbf{p}) \cdot \mathbf{v} + \mathbf{M}_{\alpha,\alpha} \cdot \boldsymbol{\omega}] \frac{\mathrm{d}A}{G},$$

where we have used (1.23). By assumption, the equations of equilibrium (1.110) are satisfied. Using this to transform the last integrand in the previous equation, we find

$$\int \mathbf{p}\cdot\mathbf{v}\,\mathrm{d}A + \oint_C (\mathbf{n}\cdot\mathbf{v} + \mathbf{m}\cdot\boldsymbol{\omega})\,\mathrm{d}s$$

$$= \int \sum_\alpha [\mathbf{N}_\alpha\cdot(\mathbf{v}_{,\alpha} - \boldsymbol{\omega}\times\mathbf{a}_\alpha) + \mathbf{M}_\alpha\cdot\boldsymbol{\omega}_{,\alpha}]\,\frac{\mathrm{d}A}{G}, \quad (1.114)$$

or

$$A_{(e)} = A_{(i)}, \quad (1.114a)$$

where the *internal work* is given by the right-hand side of (1.114). In the previous derivation, we have not as yet utilized the fact that the rotation vector is determined by the middle surface displacements. This will now be taken into account. Using (1.75) and (1.85) and introducing the formulae (1.109) for the contact forces and couples, we derive

$$\sum_\alpha [\mathbf{N}_\alpha\cdot(\mathbf{v}_{,\alpha} - \boldsymbol{\omega}\times\mathbf{a}_\alpha)] = \sum_\alpha \left(\frac{G}{A_\alpha}\left(\sum_\beta \mathrm{N}_{\alpha\beta}\mathbf{e}_\beta + \mathrm{Q}_\alpha\mathbf{e}_3\right)\cdot A_\alpha \sum_\gamma (\varepsilon_{\alpha\gamma}\mathbf{e}_\gamma)\right)$$

$$= G\sum_{\alpha,\beta} \mathrm{N}_{\alpha\beta}\varepsilon_{\alpha\beta},$$

$$\sum_\alpha \mathbf{M}_\alpha\cdot\boldsymbol{\omega}_{,\alpha} = \sum_\alpha \left(\frac{G}{A_\alpha}\sum_\beta \mathrm{M}_{\alpha\beta}\hat{\mathbf{e}}_\beta\cdot\boldsymbol{\omega}_{,\alpha}\right)$$

$$= G\sum_{\alpha,\beta} \mathrm{M}_{\alpha\beta}k_{\alpha\beta}.$$

Inserting these expressions into (1.114), we obtain

$$\int \mathbf{p}\cdot\mathbf{v}\,\mathrm{d}A + \oint_C (\mathbf{n}\cdot\mathbf{v} + \mathbf{m}\cdot\boldsymbol{\omega})\,\mathrm{d}s = \int \sum_{\alpha,\beta} (\mathrm{N}_{\alpha\beta}\varepsilon_{\alpha\beta} + \mathrm{M}_{\alpha\beta}k_{\alpha\beta})\,\mathrm{d}A. \quad (1.115)$$

We now introduce the following quantities:

$$\begin{aligned} \mathrm{N}_{(\alpha\beta)} &= \tfrac{1}{2}(\mathrm{N}_{\alpha\beta} + \mathrm{N}_{\beta\alpha}) = \mathrm{N}_{(\beta\alpha)}, \\ \mathrm{N}_{[\alpha\beta]} &= \tfrac{1}{2}(\mathrm{N}_{\alpha\beta} - \mathrm{N}_{\beta\alpha}) = -\mathrm{N}_{[\beta\alpha]}, \end{aligned} \quad (1.116)$$

so that $\mathrm{N}_{(\alpha\beta)}$ is symmetric and $\mathrm{N}_{[\alpha\beta]}$ is skew-symmetric in the indices α and β. We obviously have

$$\mathrm{N}_{\alpha\beta} = \mathrm{N}_{(\alpha\beta)} + \mathrm{N}_{[\alpha\beta]}, \quad (1.117a)$$

which provides a representation of $\mathrm{N}_{\alpha\beta}$ as the sum of a symmetric and a skew-symmetric contribution. In a similar way we put

$$\mathrm{M}_{\alpha\beta} = \mathrm{M}_{(\alpha\beta)} + \mathrm{M}_{[\alpha\beta]}, \quad (1.117b)$$

in which the quantities on the right-hand side are defined by equations which are obtained from (1.116) by replacing the letter N by M.

We now substitute (1.117) in the right-hand side of (1.115), and we also replace $k_{\alpha\beta}$ by the expression (1.97). Noting that the sum of products of corresponding components of a symmetric and a skew-symmetric set of quantities vanishes, i.e.

$$A_{\alpha\beta} = A_{\beta\alpha} \wedge B_{\alpha\beta} = -B_{\beta\alpha} \Rightarrow \sum_{\alpha,\beta} A_{\alpha\beta} B_{\alpha\beta} = 0,$$

we then deduce

$$\begin{aligned}
&\sum_{\alpha,\beta} (\mathrm{N}_{\alpha\beta}\varepsilon_{\alpha\beta} + \mathrm{M}_{\alpha\beta}k_{\alpha\beta}) \\
&= \sum_{\alpha,\beta} \left\{ (\mathrm{N}_{(\alpha\beta)} + \mathrm{N}_{[\alpha\beta]})\varepsilon_{\alpha\beta} + (\mathrm{M}_{(\alpha\beta)} + \mathrm{M}_{[\alpha\beta]}) \left[\kappa_{\alpha\beta} + \frac{1}{2}\left(\frac{1}{R_\alpha} - \frac{1}{R_\beta}\right)\varepsilon_{\alpha\beta} \right] \right\} \\
&= \sum_{\alpha,\beta} \left\{ \left[\mathrm{N}_{(\alpha\beta)} + \frac{1}{2}\left(\frac{1}{R_\alpha} - \frac{1}{R_\beta}\right)\mathrm{M}_{[\alpha\beta]} \right] \varepsilon_{\alpha\beta} + \mathrm{M}_{(\alpha\beta)}\kappa_{\alpha\beta} \right\},
\end{aligned}$$

since $\varepsilon_{\alpha\beta}$ and $\kappa_{\alpha\beta}$ are symmetric and $\frac{1}{2}(1/R_\alpha - 1/R_\beta)\varepsilon_{\alpha\beta}$ is skew-symmetric in the indices α and β. The *principle of virtual work* for the shell may therefore be expressed in the form

$$\begin{aligned}
&\int \mathbf{p} \cdot \mathbf{v} \, dA + \oint_C (\mathbf{n} \cdot \mathbf{v} + \mathbf{m} \cdot \boldsymbol{\omega}) \, ds \\
&\qquad = \int \sum_{\alpha,\beta} \left\{ \left[\mathrm{N}_{(\alpha\beta)} + \frac{1}{2}\left(\frac{1}{R_\alpha} - \frac{1}{R_\beta}\right)\mathrm{M}_{[\alpha\beta]} \right] \varepsilon_{\alpha\beta} + \mathrm{M}_{(\alpha\beta)}\kappa_{\alpha\beta} \right\} dA. \qquad (1.118)
\end{aligned}$$

We may summarize the result of our derivation in the following theorem.

Theorem 1.5.1. Suppose that the contact forces and couples satisfy the equations of equilibrium (1.111) (or the equivalent vector equations (1.110)), and that the edge loads **n**, **m** are the corresponding equilibrium values. Then the principle of virtual work (1.118) is valid for any displacement field.

We notice that the integrand on the right-hand side of (1.118), which will be called the *specific internal work*, is a sum of products of statical and geometrical quantities. The geometrical quantities are the strain and bending measures $\varepsilon_{\alpha\beta}$ and $\kappa_{\alpha\beta}$. The statical quantities are certain linear combinations of the contact forces and couples which will be called the *effective contact quantities*, and for which we introduce the notation

$$\begin{aligned}
N_{\alpha\beta} &= \mathrm{N}_{(\alpha\beta)} + \frac{1}{2}\left(\frac{1}{R_\alpha} - \frac{1}{R_\beta}\right)\mathrm{M}_{[\alpha\beta]}, \\
M_{\alpha\beta} &= \mathrm{M}_{(\alpha\beta)}.
\end{aligned} \qquad (1.119)$$

It will be seen that both $N_{\alpha\beta}$ and $M_{\alpha\beta}$ are *symmetric* in the indices α and β. This means that there are only six different effective contact quantities. Likewise, there are only six different strain and bending measures. Introducing (1.119) in (1.118) we can write the specific internal work in the form

$$\sum_{\alpha,\beta} (N_{\alpha\beta}\varepsilon_{\alpha\beta} + M_{\alpha\beta}\kappa_{\alpha\beta}). \qquad (1.120)$$

It will be recalled that we introduced ten components of the contact forces and couples in the description of the statics of the shell in section 1.5.1 (namely, $N_{\alpha\beta}$, $M_{\alpha\beta}$, and Q_α). It might be expected, therefore, that the specific internal work should be a sum of products of all the components of the contact forces and couples and a corresponding number of strain quantities (cf. the classical theory of elasticity of three-dimensional bodies, in which the specific internal work is a sum of products of all the stress components and a corresponding number of strain components). Contrary to this expectation, we have found that only six different contact forces and couple quantities and six different strain quantities appear in the expression for the specific internal work of the shell.

In order to explain the reasons for this apparent anomaly, we shall, for the moment, consider an analogous problem, namely the theory of *straight, plane beams*. It will be recalled that it is assumed in the ordinary technical theory of beams that cross-sections of the beam which are plane and perpendicular to the centre-line before deformation remain so after deformation (the so-called Bernouilli's or Navier's hypothesis). Three components of internal forces and moments are introduced in the statical description (namely, the direct force N, the shear force Q, and the bending moment M), but in the expression for the specific internal work ($N\varepsilon + M\kappa$) there appear only the two components N and M multiplied by two corresponding strain quantities (the strain ε of the centre-line and the change of curvature κ). The constitutive equations for an elastic beam, which provide the connection between the stress resultants and the strain quantities, express the statical quantities N and M which appear in the specific internal work as linear functions of ε and κ.

It is also well known that a more general theory of beams may be used, in which shearing deformations are taken into account. Cross-sections of the beam, which, in the reference state, are planes perpendicular to the centre-line, can now undergo arbitrary small rotations about axes perpendicular to the plane of the beam, i.e. they are no longer constrained to remain perpendicular to the centre-line. In this general theory, there appears an additional term $Q\varphi$ in the expression for the specific internal work, where φ is the shear strain. The specific internal work is now a sum of the products of the three stress resultants N, Q, and M and three corresponding strain quantities ε, φ, and κ. The constitutive equations of the general theory express the three stress resultants as linear functions of the three strain quantities.

It will now be seen that the technical theory of beams is a special case of the general theory, and the former theory can be obtained from the latter by introducing a *geometrical constraint*, namely that the shear strain should be zero everywhere. The corresponding stress resultant (the shear force Q) now becomes an *internal reaction* (constraint force) which makes no contribution to the internal work, because the term $Q\varphi$ in the specific internal work vanishes when $\varphi \equiv 0$. In the technical theory of beams we cannot determine the shear force by means of a constitutive equation (because the shear strain is zero everywhere), but this force must be calculated from the equilibrium condition

$$\frac{\mathrm{d}M}{\mathrm{d}x} + Q = 0.$$

We now return to the analogous problems in the theory of shells. We may regard the version of the theory of shells which has been developed in the previous sections (in the present context this will be called the *classical theory of shells*) as a special case of a more general theory of shells. In this general theory, the number of degrees of freedom used in the geometrical description of the displacements of the shell is greater than in the case of the classical theory of shells, so that a wider range of deformation patterns becomes possible for the shell. The transition from the general to the classical theory of shells corresponds to the imposition of certain geometrical constraints in the general theory. The presence of geometrical constraints manifests itself in the fact that not all of the ten components of contact forces and couples appear in the expression for the specific internal work, which only involves the six effective contact quantities, as previously explained.

Such a general theory of shells was developed by Green and by Naghdi.[28] In one of the simpler versions of this theory, the components ω_1 and ω_2 (see (1.81a)) of the rotation vector are unconstrained, i.e. independent of the middle surface displacement $\mathbf{v}$, so that the normal vector $\mathbf{a}_3$ of the middle surface in the reference state is not necessarily carried into the normal vector $\mathbf{a}_3^*$ of the deformed state as a result of the rotation (this means that shear deformations are now allowed). The geometrical constraints are expressed by equations (1.81a) and imply that the shear deformations vanish, so that the rotation vector is completely determined by the middle surface displacements as described in section 1.4.3.

We shall now show that it is possible to derive a set of equations of equilibrium which involves only the effective contact quantities, so that the internal reactions are, in fact, eliminated. It will be shown in section 1.6 that the constitutive equations of the classical theory of shells express the six effective contact quantities as unique and invertible linear functions of the six strain and bending measures. In this way we shall derive a set of governing equations for the shell in which the internal forces and moments are represented solely by the effective contact quantities.

The elimination of the internal reactions is accomplished in the following manner. It is assumed that the shell is in equilibrium under the action of the applied load $\mathbf{p}$, the contact forces and couples $\mathrm{N}_{\alpha\beta}$, $\mathrm{M}_{\alpha\beta}$, and Q_α (which will be called the *complete contact forces and couples*), and the corresponding equilibrium values $\mathbf{n}$ and $\mathbf{m}$ of the edge loads. We now introduce the quantities

$$N^{\mathrm{u}}_{\alpha\beta} = N_{\alpha\beta} + \frac{1}{2}\left(\frac{1}{R_\alpha} - \frac{1}{R_\beta}\right) M_{\alpha\beta}, \tag{1.121}$$

which depend in a simple manner on the effective contact quantities. The index u indicates that $N^{u}_{\alpha\beta}$ is generally unsymmetric in the indices α and β. Since the term

$$\frac{1}{2}\left(\frac{1}{R_\alpha}-\frac{1}{R_\beta}\right)M_{\alpha\beta}$$

is skew-symmetric in the indices α and β, we evidently have

$$N^{u}_{(\alpha\beta)}=\tfrac{1}{2}(N^{u}_{\alpha\beta}+N^{u}_{\beta\alpha})=N_{\alpha\beta},$$

so that $N^{u}_{\alpha\beta}$ differs from $N_{\alpha\beta}$ only with respect to the components N^{u}_{12} and N^{u}_{21}. We also put

$$Q_\alpha=\frac{1}{A_1A_2}[(A_\beta M_{\alpha\alpha})_{,\alpha}+(A_\alpha M_{\beta\alpha})_{,\beta}+A_{\alpha,\beta}M_{\alpha\beta}-A_{\beta,\alpha}M_{\beta\beta}], \tag{1.122}$$

where $\alpha\neq\beta$. The quantities Q_α are therefore the values of the transverse shear forces, which are obtained from the fourth and fifth equations of equilibrium $(1.111b)_{1,2}$ when the values $M_{\alpha\beta}$ are inserted for the moments. The quantities $N^{u}_{\alpha\beta}$, $M_{\alpha\beta}$, and Q_α are called the *effective contact forces and couples*. (Note that we have defined two sets of contact forces and couples: the *complete* contact forces and couples $\mathrm{N}_{\alpha\beta}$, $\mathrm{M}_{\alpha\beta}$, and Q_α (roman capitals), and the *effective* constant forces and couples $N^{u}_{\alpha\beta}$, $M_{\alpha\beta}$, and Q_α (italic capitals), (see (1.121) and (1.122)). Moreover, the term *effective contact quantities* is used to denote $N_{\alpha\beta}$ and $M_{\alpha\beta}$, where $N_{\alpha\beta}=N^{u}_{(\alpha\beta)}$.)

We now form the differences between the complete and the effective contact forces and couples. From (1.121), (1.119) and (1.117) we find

$$\Delta M_{\alpha\beta}=\mathrm{M}_{\alpha\beta}-M_{\alpha\beta}=\mathrm{M}_{\alpha\beta}-\mathrm{M}_{(\alpha\beta)}=\mathrm{M}_{[\alpha\beta]}, \tag{1.123}$$

$$\begin{aligned}\Delta N_{\alpha\beta}&=\mathrm{N}_{\alpha\beta}-N^{u}_{\alpha\beta}\\&=\mathrm{N}_{\alpha\beta}-\left[\mathrm{N}_{(\alpha\beta)}+\frac{1}{2}\left(\frac{1}{R_\alpha}-\frac{1}{R_\beta}\right)\mathrm{M}_{[\alpha\beta]}\right]-\frac{1}{2}\left(\frac{1}{R_\alpha}-\frac{1}{R_\beta}\right)\mathrm{M}_{(\alpha\beta)}\\&=\mathrm{N}_{[\alpha\beta]}-\frac{1}{2}\left(\frac{1}{R_\alpha}-\frac{1}{R_\beta}\right)\mathrm{M}_{\alpha\beta}.\end{aligned} \tag{1.124}$$

The sixth equation of equilibrium $(1.111b)_3$ can be written in the form

$$\mathrm{N}_{\alpha\beta}-\mathrm{N}_{\beta\alpha}=\mathrm{M}_{\alpha\beta}/R_\alpha-\mathrm{M}_{\beta\alpha}/R_\beta,$$

or

$$2\mathrm{N}_{[\alpha\beta]}=\mathrm{M}_{\alpha\beta}/R_\alpha-\mathrm{M}_{\beta\alpha}/R_\beta,$$

and when this is inserted in (1.124) we obtain

$$\Delta N_{\alpha\beta}=\mathrm{M}_{[\alpha\beta]}/R_\beta. \tag{1.125}$$

The skew-symmetric system $\mathrm{M}_{[\alpha\beta]}$ can always be represented in the form

$$[\mathrm{M}_{[\alpha\beta]}]=\begin{bmatrix}0 & \Phi\\ -\Phi & 0\end{bmatrix},$$

where Φ is a function of (θ_1, θ_2). When this is substituted in (1.123) and (1.125), we obtain the following expressions for the force and couple differences

$$\begin{bmatrix} \Delta N_{11} & \Delta N_{12} \\ \Delta N_{21} & \Delta N_{22} \end{bmatrix} = \begin{bmatrix} 0 & \Phi/R_2 \\ -\Phi/R_1 & 0 \end{bmatrix}, \quad \begin{bmatrix} \Delta M_{11} & \Delta M_{12} \\ \Delta M_{21} & \Delta M_{22} \end{bmatrix} = \begin{bmatrix} 0 & \Phi \\ -\Phi & 0 \end{bmatrix}. \qquad (1.126a)$$

If we put $\alpha = 1$ and $\beta = 2$ in (1.122), subtract the resulting equation from $(1.111b)_1$ and use (1.126a) we get

$$\begin{aligned} \Delta Q_1 = \mathrm{Q}_1 - Q_1 &= \frac{1}{A_1A_2}[(A_1\Delta M_{21})_{,2} + A_{1,2}\Delta M_{12}] \\ &= \frac{1}{A_1A_2}[-(A_1\Phi)_{,2} + A_{1,2}\Phi] \\ &= -\Phi_{,2}/A_2. \end{aligned}$$

ΔQ_2 can be found in a similar manner, so that we obtain for the shear force differences

$$[\Delta Q_1 \quad \Delta Q_2] = [-\Phi_{,2}/A_2 \quad \Phi_{,1}/A_1]. \qquad (1.126b)$$

We now prove that the force and couple differences (1.126) satisfy the homogeneous equations of equilibrium, i.e. (1.111) with $p_i = 0$. It is immediately obvious that the fourth and fifth equations $(1.111b)_{1,2}$ are satisfied, since these equations were used to define ΔQ_α. By substituting (1.126) in the remaining equations of (1.111) we find, using $(1.111a)_1$:

$$\begin{aligned} &-(A_1\Phi/R_1)_{,2} + A_{1,2}\Phi/R_2 + A_1\Phi_{,2}/R_1 \\ &\qquad = -A_{1,2}\Phi/R_2 - A_1\Phi_{,2}/R_1 + A_{1,2}\Phi/R_2 + A_1\Phi_{,2}/R_1 = 0, \end{aligned}$$

in which we have used Codazzi's equation $(1.43)_1$. The second equation $(1.111a)_2$ is treated in a similar manner. Using $(1.111a)_3$ we get:

$$-(A_2\Phi_{,2}/A_2)_{,1} + (A_1\Phi_{,1}/A_1)_{,2} = 0.$$

And finally using the last equation $(1.111b)_3$, we find:

$$\Phi/R_2 + \Phi/R_1 - \Phi/R_1 - \Phi/R_2 = 0.$$

This completes the proof.

We now return to the effective contact forces and couples (1.121) and (1.122). Since the complete contact forces and couples are in equilibrium with the actual applied surface load **p**, and the force and couple differences are in equilibrium with the surface load $\mathbf{p} = \mathbf{0}$, we conclude that the effective contact forces $N^{\mathrm{u}}_{\alpha\beta} = \mathrm{N}_{\alpha\beta} - \Delta N_{\alpha\beta}$, etc., are in equilibrium with the applied surface load **p** (this follows from the fact that the equilibrium conditions (1.111) are linear in the contact forces and couples). We have therefore proved the following theorem.

Theorem 1.5.2. The effective contact forces and couples are in equilibrium with the applied surface load **p**. If we therefore replace the complete contact forces and couples by the effective contact forces and couples in the equations of equilibrium (1.111), the equations remain valid.

Since the effective contact forces and couples can be expressed in terms of the effective contact quantities (see (1.121) and (1.122)), we have therefore arrived at a set of equilibrium conditions in which the internal forces and moments appear solely in the form of the six effective contact quantities. Since the latter are independent of the indeterminate function Φ, it follows that we have, in fact, eliminated the internal reactions. Substituting (1.120) in (1.118) we finally obtain

$$\int \mathbf{p} \cdot \mathbf{v} \, \mathrm{d}A + \oint_C (\mathbf{n} \cdot \mathbf{v} + \mathbf{m} \cdot \boldsymbol{\omega}) \, \mathrm{d}s = \int \sum_{\alpha,\beta} (N_{\alpha\beta}\varepsilon_{\alpha\beta} + M_{\alpha\beta}\kappa_{\alpha\beta}) \, \mathrm{d}A, \quad (1.127)$$

which is a statement of the principle of virtual work with the internal work expressed in terms of the effective contact quantities.

It was mentioned previously that the contact forces and couples determined by solving a problem in the theory of shells are, in fact, the effective contact quantities (and thus the effective contact forces and couples $N^{u}_{\alpha\beta}$, $M_{\alpha\beta}$, and Q_α (see (1.121) and (1.122))). It may now be asked to what extent knowledge of the effective contact forces and couples will enable us to determine the complete contact forces and couples. It follows from our previous investigation that, if the effective contact forces and couples are given, the corresponding values of the complete contact forces and couples cannot be determined uniquely. If we substitute any differentiable function Φ on the right-hand side of the formulae (1.126) for the force and couple differences, then the sum of these and the given effective contact forces and couples will constitute a possible system of complete contact forces and couples. This proves the lack of uniqueness.

This arbitrariness has, however, no practical significance for the applicability of the theory of shells. It can be shown by comparison with results from the three-dimensional theory that the deviations between the values of the complete contact forces and couples computed from the results of the three-dimensional theory and those determined by means of the theory of shells is insignificant in the major part of a thin shell apart from a narrow zone along the boundary curve.

1.5.4 Static–geometric analogy, stress functions

We shall now consider an interesting analogy between statical and geometrical quantities which may be used to express the effective contact forces and couples in terms of three stress functions.

We begin by writing the homogeneous equations of equilibrium (i.e. (1.110) with $\mathbf{p} = \mathbf{0}$) and the compatibility conditions (1.101), and (1.101a) side by side:

$$\mathbf{N}_{1,1} + \mathbf{N}_{2,2} = \mathbf{0}, \qquad \mathbf{k}_{2,1} + (-\mathbf{k}_1)_{,2} = \mathbf{0},$$
$$\mathbf{M}_{1,1} + \mathbf{M}_{2,2} + \mathbf{a}_1 \times \mathbf{N}_1 + \mathbf{a}_2 \times \mathbf{N}_2 = \mathbf{0}, \qquad \boldsymbol{\gamma}_{2,1} + (-\boldsymbol{\gamma}_1)_{,2} + \mathbf{a}_1 \times \mathbf{k}_2 + \mathbf{a}_2 \times (-\mathbf{k}_1) = \mathbf{0}.$$

It will now be seen that the two sets of equations have the same form, so that the statical quantities $\mathbf{N}_1$, $\mathbf{N}_2$, $\mathbf{M}_1$, and $\mathbf{M}_2$ correspond to the geometrical quantities $\mathbf{k}_2$, $-\mathbf{k}_1$, $\boldsymbol{\gamma}_2$ and $-\boldsymbol{\gamma}_1$, respectively. We also note that the moment vectors $\mathbf{M}_1$, $\mathbf{M}_2$ and the corresponding geometrical quantities $\boldsymbol{\gamma}_2$, $-\boldsymbol{\gamma}_1$ are vectors parallel to the tangent plane, while the force vectors $\mathbf{N}_1$, $\mathbf{N}_2$ and the corresponding geometrical quantities $\mathbf{k}_2$, $-\mathbf{k}_1$ may have components in the directions of all three unit vectors $\mathbf{e}_i$. The following analogy can therefore be established between static and geometric quantities:

$\mathbf{N}_1, \mathbf{N}_2$	correspond to	$\mathbf{k}_2, -\mathbf{k}_1$
$\mathbf{M}_1, \mathbf{M}_2$		$\boldsymbol{\gamma}_2, -\boldsymbol{\gamma}_1$
$\begin{bmatrix} \mathrm{N}_{11} & \mathrm{N}_{12} \\ \mathrm{N}_{21} & \mathrm{N}_{22} \end{bmatrix}$		$\begin{bmatrix} -k_{22} & k_{21} \\ k_{12} & -k_{11} \end{bmatrix}$
$A_2\mathrm{Q}_1, A_1\mathrm{Q}_2$		$k_2, -k_1$
$\begin{bmatrix} \mathrm{M}_{11} & \mathrm{M}_{12} \\ \mathrm{M}_{21} & \mathrm{M}_{22} \end{bmatrix}$		$\begin{bmatrix} \varepsilon_{22} & -\varepsilon_{21} \\ -\varepsilon_{12} & \varepsilon_{11} \end{bmatrix}$

in which the analogy between the components is obtained by a comparison between the corresponding vectors in the first two lines and application of (1.109) and the equations

$$\begin{aligned} \boldsymbol{\gamma}_\alpha &= A_\alpha \sum_\beta \varepsilon_{\alpha\beta}\mathbf{e}_\beta, \\ \mathbf{k}_\alpha &= A_\alpha \sum_\beta k_{\alpha\beta}\hat{\mathbf{e}}_\beta + k_\alpha \mathbf{e}_3 \end{aligned} \tag{1.128}$$

(see (1.99)).

The static–geometric analogy may be formulated in the following manner. If we replace the statical quantities by the corresponding geometrical quantities in the homogeneous equations of equilibrium, then the resulting equations become identical with the compatibility conditions.

We now turn our attention to the problem of stress functions. It follows from the static–geometric analogy that any set of contact forces and couples $\mathrm{N}'_{\alpha\beta}$, $A_2\mathrm{Q}'_1$, $A_1\mathrm{Q}'_2$, $\mathrm{M}'_{\alpha\beta}$ which satisfies the homogeneous equations of equilibrium (i.e. (1.111) with $p_i = 0$) and for which $\mathrm{M}'_{12} = \mathrm{M}'_{21}$, with the help of the analogy can be regarded as a set of strain and bending measures $k'_{\nu\mu}$, k'_ν, $\varepsilon'_{\nu\mu}$ which satisfies the compatibility conditions $(1.101)_1$ and (1.101a). (Note that the condition $\mathrm{M}'_{12} = \mathrm{M}'_{21}$ must be imposed in order to ensure the validity of the equation $\varepsilon'_{21} = \varepsilon'_{12}$, which must always be satisfied by the corresponding geometric quantities.)

It was shown in section 1.4.5 that when the compatibility conditions are satisfied, then a displacement field $\mathbf{v}'$ exists, for which $k'_{\nu\mu}$, k'_ν, and $\varepsilon'_{\nu\mu}$ are the corresponding deformation measures. This means, however, that the corresponding contact forces and couples $N'_{\alpha\beta}$, A_2Q_1, A_1Q_2, and $M'_{\alpha\beta}$ can be expressed in terms of three stress functions (namely, the components of the vector $\mathbf{v}'$) in the same way as $k'_{\nu\mu}$, k'_{ν}, and $\varepsilon'_{\nu\mu}$ are expressed as functions of the displacement components.

Since the effective contact forces and couples satisfy the symmetry condition $M_{12} = M_{21}$ which is the characteristic property of the above contact forces and couples $N'_{\alpha\beta}, \ldots$, the result of our analysis may be stated in the following manner.

Any set of *effective contact forces and couples* which satisfies the *homogeneous equations of equilibrium* can be represented in terms of *three stress functions* $\varphi_i(\theta_1, \theta_2)$, $i = 1, 2, 3$, in the following manner:

$$\begin{aligned}
N^{\mathrm{u}}_{11} &= -\kappa_{22}(\varphi_1, \varphi_2, \varphi_3),\\
N^{\mathrm{u}}_{22} &= -\kappa_{11}(\varphi_1, \varphi_2, \varphi_3),\\
N^{\mathrm{u}}_{12} &= \kappa_{12}(\varphi_1, \varphi_2, \varphi_3) + \frac{1}{2}\left(\frac{1}{R_2} - \frac{1}{R_1}\right)\varepsilon_{12}(\varphi_1, \varphi_2, \varphi_3),\\
N^{\mathrm{u}}_{21} &= \kappa_{12}(\varphi_1, \varphi_2, \varphi_3) - \frac{1}{2}\left(\frac{1}{R_2} - \frac{1}{R_1}\right)\varepsilon_{12}(\varphi_1, \varphi_2, \varphi_3),\\
M_{11} &= \varepsilon_{22}(\varphi_1, \varphi_2, \varphi_3),\\
M_{22} &= \varepsilon_{11}(\varphi_1, \varphi_2, \varphi_3),\\
M_{12} &= M_{21} = -\varepsilon_{12}(\varphi_1, \varphi_2, \varphi_3),
\end{aligned} \tag{1.129}$$

the corresponding transverse shear forces Q_α being given by (1.122). $\varepsilon_{\alpha\beta}(\varphi_1, \varphi_2, \varphi_3)$ and $\kappa_{\alpha\beta}(\varphi_1, \varphi_2, \varphi_3)$ denote the expressions (1.66) and (1.96) for the strain and bending measures with v_i replaced by φ_i. It should also be noted that we have used (1.97) to express $k_{\nu\mu}$ in terms of $\kappa_{\nu\mu}$ and $\varepsilon_{\nu\mu}$ in the formulae for $N^{\mathrm{u}}_{\alpha\beta}$.

1.6 The strain energy of the shell, constitutive equations

In the preceding sections we have described the deformation of the shell by means of the strain measures $\varepsilon_{\alpha\beta}$ and the bending measures $\kappa_{\alpha\beta}$, and we have formulated the equilibrium conditions for the shell by means of the effective contact quantities. The connection between these geometrical and statical quantities is furnished by the constitutive equations.

A detailed derivation of the constitutive equations is beyond the scope of the present treatment of the theory of shells. We shall confine ourselves to some introductory remarks followed by a statement of the expression for the strain energy of the shell, which is then used to deduce the constitutive equations.

We assume that *the shell is elastic*, and in analogy with the three-dimensional case we also assume that the *specific strain energy of the shell* (i.e.

the strain energy per unit area of the middle surface) is a positive definite quadratic form in the strain and bending measures (i.e. in the quantities that determine the deformation of the shell). Let us introduce the vectors

$$\boldsymbol{\varepsilon}^{\mathrm{T}} = [\varepsilon_{11} \quad \varepsilon_{22} \quad \varepsilon_{12}], \qquad \boldsymbol{\kappa}^{\mathrm{T}} = [\kappa_{11} \quad \kappa_{22} \quad \kappa_{12}],$$

where the index T denotes transposition. It now follows from the above assumptions that the specific strain energy can be written in the form

$$W = \tfrac{1}{2}[\boldsymbol{\varepsilon}^{\mathrm{T}}\ \boldsymbol{\kappa}^{\mathrm{T}}]\mathbf{D}\begin{bmatrix}\boldsymbol{\varepsilon}\\ \boldsymbol{\kappa}\end{bmatrix}, \tag{1.130}$$

in which $\mathbf{D}$ is a (6, 6) symmetric, positive definite matrix, the elements of which depend on the geometric properties of the shell (i.e. the principal curvatures and the shell thickness) and on parameters that characterize the material properties of the shell. If we write $\mathbf{D}$ in the form

$$\mathbf{D} = \begin{bmatrix}\mathbf{D}_{11} & \mathbf{D}_{12}\\ \mathbf{D}_{12}^{\mathrm{T}} & \mathbf{D}_{22}\end{bmatrix},$$

where the submatrices $\mathbf{D}_{\alpha\beta}$ are of order (3, 3) and the matrices $\mathbf{D}_{\alpha\alpha}$ are symmetrical, the specific strain energy can be expressed in the form

$$W = \tfrac{1}{2}[\boldsymbol{\varepsilon}^{\mathrm{T}}\mathbf{D}_{11}\boldsymbol{\varepsilon} + \boldsymbol{\kappa}^{\mathrm{T}}\mathbf{D}_{22}\boldsymbol{\kappa} + 2\boldsymbol{\varepsilon}^{\mathrm{T}}\mathbf{D}_{12}\boldsymbol{\kappa}]. \tag{1.131}$$

It will be seen that the first and second terms on the right-hand side are quadratic forms in the strain and bending measures, respectively, while the third term contains products of strain and bending measures and represents a coupling of stretching and bending effects.

If the shell is regarded as a three-dimensional body, and the classical theory of elasticity, together with the Love–Kirchhoff hypothesis, is employed, it can be shown that the specific strain energy is given by an expression of the form (1.131). Moreover, in the case of thin shells it can be shown that the coupling term may be omitted, since the contribution from this term is smaller than the error that is introduced in any case because of the use of the Love–Kirchhoff hypothesis (see for example Koiter's investigation[22]).

For a *thin shell* consisting of a homogeneous, isotropic, elastic material it is found that the expression for the *specific strain energy of the shell* is given by

$$\begin{aligned} W = \frac{1}{2}\,&\frac{Eh}{(1-\nu^2)}[(1-\nu)(\varepsilon_{11}^2 + \varepsilon_{22}^2 + 2\varepsilon_{12}^2) + \nu(\varepsilon_{11} + \varepsilon_{22})^2] \\ &+ \frac{Eh^3}{24(1-\nu^2)}[(1-\nu)(\kappa_{11}^2 + \kappa_{22}^2 + 2\kappa_{12}^2) + \nu(\kappa_{11} + \kappa_{22})^2], \end{aligned} \tag{1.132}$$

in which E is Young's modulus, ν is Poisson's ratio, and h is the thickness of the shell. It is well known that the inequality $0 \leqslant \nu \leqslant \frac{1}{2}$ holds for Poisson's ratio. It follows from this and (1.132) that W is positive definite, i.e. $W \geqslant 0$, where the equality sign applies only when all the strain and bending measures vanish. In the literature on shells the expression (1.132) is often called Love's first approximation for the specific strain energy.

It can be shown[25,26] that the expression $\int W\,\mathrm{d}A$ for the total strain energy of the shell (with W given by (1.132)) approximates the total strain energy of the corresponding three-dimensional body according to the classical theory of elasticity with a relative error of the order

$$\left(\frac{h}{L}\right)^2 + \frac{h}{R}\,.$$

In this expression R denotes the numerically smallest principal radius of curvature of the middle surface, and L denotes the smallest 'wavelength' of the deformation pattern on the middle surface. L is defined by

$$\left|\frac{\mathrm{d}\varepsilon}{\mathrm{d}s}\right| = O\left(\frac{\varepsilon}{L}\right) \wedge \left|\frac{\mathrm{d}\kappa}{\mathrm{d}s}\right| = O\left(\frac{\kappa}{L}\right), \tag{1.132a}$$

where s is the arc length of any curve on the middle surface, and ε and κ are the numerically largest principal values of the middle surface strain and bending measures, respectively.

In order to arrive at a set of constitutive equations we shall use the following argument. In analogy with the relations in the three-dimensional theory of elasticity, we shall assume that the specific internal work (1.120) for infinitesimal increments of the strain and bending measures is equal to the differential of the specific strain energy, i.e.

$$\sum_{\alpha,\beta} (N_{\alpha\beta}\,\mathrm{d}\varepsilon_{\alpha\beta} + M_{\alpha\beta}\,\mathrm{d}\kappa_{\alpha\beta}) = \mathrm{d}W, \tag{1.133}$$

or, written in full,

$$\begin{aligned} N_{11}\,\mathrm{d}\varepsilon_{11} + N_{22}\,\mathrm{d}\varepsilon_{22} + 2N_{12}\,\mathrm{d}\varepsilon_{12} + M_{11}\,\mathrm{d}\kappa_{11} + M_{22}\,\mathrm{d}\kappa_{22} + 2M_{12}\,\mathrm{d}\kappa_{12} \\ = \frac{\partial W}{\partial \varepsilon_{11}}\mathrm{d}\varepsilon_{11} + \frac{\partial W}{\partial \varepsilon_{22}}\mathrm{d}\varepsilon_{22} + \frac{\partial W}{\partial \varepsilon_{12}}\mathrm{d}\varepsilon_{12} + \frac{\partial W}{\partial \kappa_{11}}\mathrm{d}\kappa_{11} \\ + \frac{\partial W}{\partial \kappa_{22}}\mathrm{d}\kappa_{22} + \frac{\partial W}{\partial \kappa_{12}}\mathrm{d}\kappa_{12}. \end{aligned}$$

This equation is valid for arbitrary values of the differentials $\mathrm{d}\varepsilon_{\alpha\beta}$ and $\mathrm{d}\kappa_{\alpha\beta}$. If we in turn put one of these differentials equal to a non-zero constant and the rest equal to zero, we deduce the following relations

$$\begin{aligned} N_{\alpha\alpha} &= \partial W/\partial \varepsilon_{\alpha\alpha}, \qquad & N_{12} = N_{21} &= \tfrac{1}{2}\partial W/\partial \varepsilon_{12}, \\ M_{\alpha\alpha} &= \partial W/\partial \kappa_{\alpha\alpha}, \qquad & M_{12} = M_{21} &= \tfrac{1}{2}\partial W/\partial \kappa_{12}. \end{aligned} \tag{1.134}$$

Evaluating the partial derivatives of W by means of (1.132) and substituting the result in (1.134), we get

$$
\begin{aligned}
N_{11} &= \frac{Eh}{1-\nu^2}(\varepsilon_{11}+\nu\varepsilon_{22}),\\
N_{22} &= \frac{Eh}{1-\nu^2}(\varepsilon_{22}+\nu\varepsilon_{11}),\\
N_{12} = N_{21} &= \frac{Eh}{1+\nu}\varepsilon_{12},\\
M_{12} = M_{21} &= \frac{Eh^3}{12(1+\nu)}\kappa_{12},\\
M_{11} &= \frac{Eh^3}{12(1-\nu^2)}(\kappa_{11}+\nu\kappa_{22}),\\
M_{22} &= \frac{Eh^3}{12(1-\nu^2)}(\kappa_{22}+\nu\kappa_{11}).
\end{aligned}
\tag{1.135}
$$

These are the *constitutive equations* for the six *effective contact quantities*, which are expressed as linear functions of the six strain and bending measures. We note that the constitutive equations are uncoupled since $N_{\alpha\beta}$ depends only on $\varepsilon_{\alpha\beta}$, and $M_{\alpha\beta}$ depends only on $\kappa_{\alpha\beta}$. The equations (1.135) may be written more briefly in the form

$$
\begin{aligned}
\mathrm{N}_{\alpha\beta} &= \frac{Eh}{1-\nu^2}[(1-\nu)\varepsilon_{\alpha\beta}+\nu\delta_{\alpha\beta}(\varepsilon_{11}+\varepsilon_{22})],\\
M_{\alpha\beta} &= \frac{Eh^3}{12(1-\nu^2)}[(1-\nu)\kappa_{\alpha\beta}+\nu\delta_{\alpha\beta}(\kappa_{11}+\kappa_{22})],
\end{aligned}
\tag{1.136}
$$

where $\delta_{\alpha\beta}$ is the Kronecker delta (see (1.93)). It is easily verified that (1.135) can be solved uniquely with respect to $\varepsilon_{\alpha\beta}$ and $\kappa_{\alpha\beta}$, i.e. the strain and bending measures can be expressed as linear functions of the effective contact quantities.

Constitutive equations for the *effective contact forces and couples* are obtained by inserting (1.135) in (1.121). Thus

$$
\begin{aligned}
N^{\mathrm{u}}_{11} &= \frac{Eh}{1-\nu^2}(\varepsilon_{11}+\nu\varepsilon_{22}),\\
N^{\mathrm{u}}_{22} &= \frac{Eh}{1-\nu^2}(\varepsilon_{22}+\nu\varepsilon_{11}),\\
N^{\mathrm{u}}_{12} &= \frac{Eh}{1+\nu}\left[\varepsilon_{12}+\left(\frac{1}{R_1}-\frac{1}{R_2}\right)\frac{h^2}{24}\kappa_{12}\right],\\
N^{\mathrm{u}}_{21} &= \frac{Eh}{1+\nu}\left[\varepsilon_{12}-\left(\frac{1}{R_1}-\frac{1}{R_2}\right)\frac{h^2}{24}\kappa_{12}\right],
\end{aligned}
\tag{1.137}
$$

and the effective contact couples $M_{\alpha\beta}$ and transverse shear forces Q_α are given by $(1.135)_{4,5,6}$ and (1.122), respectively.

We shall finally consider the special case of a plane plate, for which $1/R_1 = 1/R_2 = 0$. It has been mentioned previously that the strain and bending measures in this case reduce to the usual expressions for plates, and we now see that (1.135) reduces to the well known constitutive equations for the stretching and bending of plane elastic plates.

1.7 Edge conditions

In the previous sections we have derived a number of equations which must be satisfied everywhere on the middle surface. These equations comprise the formulae (1.66) and (1.96) for the strain and bending measures in terms of the displacements, the equations of equilibrium (1.111) expressed in terms of the effective contact forces and couples, and the constitutive equations (1.137). However, a complete formulation of the problem also requires that appropriate edge conditions be specified along the boundary curve C.

For the sake of simplicity, we shall here restrict ourselves to the case in which the boundary curve C is made up of a number of arcs, each of which is part of a coordinate curve (many of the problems encountered in practical applications belong to this category).

We therefore consider an *arc of a boundary curve* which joins the points P and Q and is *part of a θ_2-curve*. The shell is under the action of *applied edge loads* **n** and **m** along the boundary curve, and these are in equilibrium with the complete contact forces and couples in the manner indicated by equation (1.107). Using (1.105) and (1.108) we then find the following equilibrium conditions at the boundary arc:

$$\begin{aligned} \mathbf{n} = \mathbf{n}_1 &= \mathrm{N}_{11}\mathbf{e}_1 + \mathrm{N}_{12}\mathbf{e}_2 + \mathrm{Q}_1\mathbf{e}_3, \\ \mathbf{m} = \mathbf{m}_1 &= \mathrm{M}_{11}\hat{\mathbf{e}}_1 + \mathrm{M}_{12}\hat{\mathbf{e}}_2. \end{aligned} \tag{1.138}$$

This means that, for the complete contact forces and couples, *each of the five components* N_{11}, N_{12}, Q_1, M_{11}, *and* M_{12} *at the boundary arc* PQ *is equal to the corresponding component of the applied edge load.*

It was shown in section 1.5.3 that the differences between the complete and the effective contact forces and couples have the form (1.126), so that, for the values at the boundary arc PQ, we have

$$\begin{aligned} \mathrm{N}_{11} &= N^{\mathrm{u}}_{11}, \qquad \mathrm{N}_{12} = N^{\mathrm{u}}_{12} + \Phi/R_2, \\ \mathrm{Q}_1 &= Q_1 - \phi_{,2}/A_2, \\ \mathrm{M}_{11} &= M_{11}, \qquad \mathrm{M}_{12} = M_{12} + \Phi. \end{aligned} \tag{1.139}$$

In these equations $\Phi = \frac{1}{2}(\mathrm{M}_{12} - \mathrm{M}_{21})$ is an internal reaction that cannot be determined within the framework of shell theory (see section 1.5.3). If the five components of the edge loads (i.e. N_{11}, N_{12}, Q_1, M_{11}, and M_{12}) are given, we cannot determine the corresponding components of the effective contact forces and couples in a unique manner by means of equations (1.139) on account of the contributions from the indeterminate function Φ. We can, however, eliminate the function Φ from the equations (1.139) and thereby derive a set of relations that is independent of Φ. From the last of equations (1.139) we find $\Phi = \mathrm{M}_{12} - M_{12}$, and when this is substituted in the remaining equations we get

$$\begin{aligned} &\mathrm{N}_{11} = N^{\mathrm{u}}_{11}, \qquad \mathrm{N}_{12} - \mathrm{M}_{12}/R_2 = N^{\mathrm{u}}_{12} - M_{12}/R_2, \\ &\mathrm{Q}_1 + \mathrm{M}_{12,2}/A_2 = Q_1 + M_{12,2}/A_2, \\ &\mathrm{M}_{11} = M_{11}. \end{aligned} \tag{1.140}$$

The four quantities

$$\mathrm{N}_{11}, \qquad (\mathrm{N}_{12} - \mathrm{M}_{12}/R_2), \qquad (\mathrm{Q}_1 + \mathrm{M}_{12,2}/A_2), \qquad \mathrm{M}_{11} \tag{1.141}$$

are called the *reduced edge loads*. It follows from the derivation of (1.140) that they are independent of the internal reaction Φ.

We may derive four equations similar to (1.140) by considering an arc of the boundary curve which is *part of a θ_1-curve*. These equations can be obtained from (1.140) by interchanging the indices 1 and 2.

We shall now consider a *corner point* Q of the boundary curve at which an arc PQ of a θ_2-curve intersects an arc QR of a θ_1-curve (see Fig. 18). As these curves are parts of coordinate curves, they intersect at right-angles at point Q. From (1.123) and (1.126) we now find

$$\begin{aligned} (\mathrm{M}_{12})_{\mathrm{Q}} &= (M_{12} + \Phi)_{\mathrm{Q}}, \\ (\mathrm{M}_{21})_{\mathrm{Q}} &= (M_{21} - \Phi)_{\mathrm{Q}}, \end{aligned} \tag{1.142}$$

and by addition, since $M_{12} = M_{21}$,

$$(\mathrm{M}_{12} + \mathrm{M}_{21})_{\mathrm{Q}} = 2(M_{12})_{\mathrm{Q}}. \tag{1.143}$$

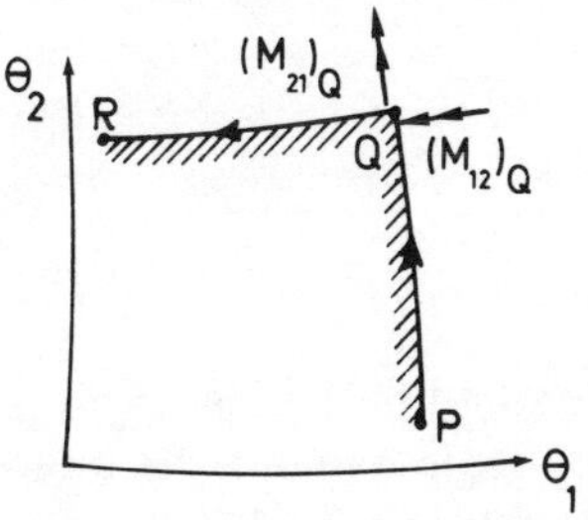

Fig. 18. Corner point on boundary curve

In this equation, $(\mathrm{M}_{12})_{\mathrm{Q}}$ and $(\mathrm{M}_{21})_{\mathrm{Q}}$ denote the values of the applied torsional moments immediately before and after the corner (as explained in connection with equation (1.138), these quantities also equal the values of the complete contact couples M_{12} and M_{21} at point Q). It follows from the derivation of (1.143) that the quantity $(M_{12} + M_{21})$ at the corner point Q is independent of the internal reaction Φ.

We now show that the work of the edge loads (i.e. the line integral on the left-hand side of (1.118)) can be transformed in such a way that the static quantities in the resulting work expression become the reduced edge loads and the quantities $(M_{12} + M_{21})$ at the corner point. Using (1.83), (1.105) and (1.108), we find for the contribution to the line integral from the arc PQ

$$\oint_P^Q (\mathbf{n}_1 \cdot \mathbf{v} + \mathbf{m}_1 \cdot \boldsymbol{\omega})\, ds_2$$

$$= \oint_P^Q (N_{11}\nu_1 + N_{12}\nu_2 + Q_1\nu_3 + M_{11}\omega_1 + M_{12}\omega_2)\, ds_2$$

$$= \oint_P^Q [N_{11}\nu_1 + N_{12}\nu_2 + Q_1\nu_3 + M_{11}\omega_1 - M_{12}(\nu_{3,2}/A_2 + \nu_2/R_2)]\, ds_2. \quad (1.144)$$

Integrating by parts the term $-M_{12}\nu_{3,2}/A_2$ in the integrand, we get

$$\oint_P^Q (\mathbf{n}_1 \cdot \mathbf{v} + \mathbf{m}_1 \cdot \boldsymbol{\omega})\, ds_2$$

$$= \oint_P^Q [N_{11}\nu_1 + (N_{12} - M_{12}/R_2)\nu_2 + (Q_1 + M_{12,2}/A_2)\nu_3 + M_{11}\omega_1]\, ds_2 - [M_{12}\nu_3] \quad (1.145)$$

It will now be seen that the reduced edge loads (1.141) appear as coefficients to the displacement quantities ν_1, ν_2, ν_3, and ω_1 in the integral on the right-hand side of (1.145).

We now apply a similar transformation to the following boundary arc and note that the positive direction of the boundary curve (see Fig. 18) is opposite to the positive θ_1-direction on the arc QR. Hence

$$\int_Q^R (\mathbf{n} \cdot \mathbf{v} + \mathbf{m} \cdot \boldsymbol{\omega})\, dx = \int_Q^R (\mathbf{n}_2 \cdot \mathbf{v} + \mathbf{m}_2 \cdot \boldsymbol{\omega})(-ds_1)$$

$$= \int_R^Q (\mathbf{n}_2 \cdot \mathbf{v} + \mathbf{m}_2 \cdot \boldsymbol{\omega})\, dx_1,$$

and it follows that the integrand in the last expression can be obtained from the integrand in (1.144) by interchanging the indices 1 and 2, so that a term $-[M_{21}\nu_3]_R^Q$ arises which, together with the corresponding term from the arc PQ, gives a total contribution at the corner point

$$-[(M_{12} + M_{21})\nu_3]_Q, \quad (1.145a)$$

so that the coefficient of ν_3 is precisely $(M_{12} + M_{21})_Q$. This proves the above assertion.

Let us now assume that (a) a set of effective contact forces and couples and (b) the applied edge loads along the boundary curve are given. We shall prove that if the conditions (1.140) and (1.143) are satisfied, then a corresponding system of complete contact forces and couples can always be determined whose values on the boundary curve are identical with the given edge loads.

When the conditions (1.140) are satisfied by the given effective contact forces and couples and the given edge loads, the values of the function Φ on the edge curve can be determined in the following manner. From $(1.139)_5$ we find $\Phi = \mathrm{M}_{12} - M_{12}$, and when this expression for Φ is inserted in the formulae (1.139) for N_{12} and Q_1 we obtain the second and third equations $(1.140)_{2,3}$ which are satisfied by assumption. We finally show that the condition (1.143) ensures the continuity of the function Φ at the corner point. The values of Φ before and after the corner point Q are given by (see (1.142) and Fig. 18)

$$\Phi_{\text{before}} = \mathrm{M}_{12} - M_{12}, \qquad \Phi_{\text{after}} = M_{12} - \mathrm{M}_{21},$$

and application of (1.143) then shows that $\Phi_{\text{before}} = \Phi_{\text{after}}$. The values of Φ have now been determined on the boundary curve, and any continuous function which assumes these values on the boundary curve can then be used to form a system of force and couple differences (1.126) and thus a system of complete contact forces and couples which corresponds to the given effective contact forces and couples.

It was mentioned previously that the statical quantities that are determined when we solve the governing equations of the theory of shells are in fact a set of effective contact forces and couples. All systems of complete contact forces and couples which correspond to this set of effective contact forces and couples (and which therefore differ from it by terms of the form (1.126)) are thus equivalent in the sense that they correspond to the same solution of the governing equations. It follows that if the applied edge loads are altered in such a manner that the reduced edge loads remain unchanged, then the solution is not affected because such an alteration can always be produced by an appropriate choice of the function Φ.

The results obtained may be summed up as follows. In the classical theory of shells, the necessary and sufficient conditions for equilibrium at the boundary arc are given by the four equations (1.140), in which the quantities on the left-hand sides of the equations should be interpreted as the components of the applied edge loads. The equilibrium conditions at the boundary are therefore expressed with the help of the reduced edge loads.

It can be shown that if the boundary curve is divided into a finite number of arcs, and *four edge conditions* of appropriate type are specified along each of these arcs (supplemented in certain cases by corner conditions of the type (1.143)), then the solution of the governing equations of the theory of shells is *unique*. The edge conditions may be geometrical (i.e. the displacements and rotations are subjected to certain constraints along the boundary curve), or they may be statical (i.e. they express equilibrium conditions at the boundary curve). Statical edge conditions are formulated with the help of the reduced edge loads (1.140).

We now give some examples of the kind of edge conditions that may be imposed along the boundary curve. The conditions will be formulated for a boundary arc PQ which is part of a θ_2-curve.

(a) *Clamped edge*. The displacements v_i and the rotation ω_1 vanish along the boundary curve. Hence, by (1.83),

$$v_1 = 0, \qquad v_2 = 0, \qquad v_3 = 0, \qquad v_{3,1} = 0 \tag{1.146}$$

on the boundary arc. It will be seen that in this case the four edge conditions are all geometrical.

(b) *Simply supported edge*. In this case the displacements v_i and the applied bending moment M_{11} are zero. Thus, by (1.140),

$$v_1 = 0, \qquad v_2 = 0, \qquad v_3 = 0, \qquad M_{11} = 0 \tag{1.147}$$

on the boundary arc. It will be seen that three of the edge conditions are geometrical and one is statical.

(c) *Free edge*. In this case the applied edge loads are zero. According to (1.140) this implies that the reduced edge loads vanish, i.e.

$$\begin{aligned} &N^{\mathrm{u}}_{11} = 0, && N^{\mathrm{u}}_{12} - M_{12}/R_2 = 0, \\ &Q_1 + M_{12,2}/A_2 = 0, && M_{11} = 0 \end{aligned} \tag{1.148}$$

on the boundary arc. In the case of free corner points this is supplemented by conditions of the type (see (1.143))

$$M_{12} = 0 \tag{1.148a}$$

at the corner points. All the four edge conditions are here statical.

1.8 Discussion of governing equations

The governing equations of the theory of shells which were derived in the preceding sections may be classified in the following manner:

(1) the strain and bending measures expressed in terms of the displacements (equations (1.66) and (1.96));

(2) the equations of equilibrium (1.111) written in terms of the effective contact forces and couples (i.e. with N_{11}, N_{12}, etc., replaced by N^{u}_{11}, N^{u}_{12}, etc.);

(3) the constitutive equations (1.137) which express the effective contact forces and couples as functions of the strain and bending measures;

(4) the equations of compatibility (1.104);

(5) the equations (1.129) which express any system of effective contact forces ar.d couples that satisfy the homogeneous equations of equilibrium in terms of three stress functions;

(6) the edge conditions (see section 1.7).

In the same way as in the classical theory of elasticity of three-dimensional bodies, two alternative methods of solution may be used in the theory of shells. We may derive a set of equations either for determination of the displacements (method A) or for determination of the contact forces and couples (method B).

Method A. Consider the equations of equilibrium (1.111) written in terms of the effective contact forces and couples. The fourth and fifth equations are solved with respect to the transverse shear forces Q_α, and these expressions for Q_α are substituted in the first three equations. In these three equations we then express the effective contact forces and couples in terms of the strain and bending measures by means of the constitutive equations, and we finally express the strain and bending measures as functions of the displacements. In this way we arrive at a system of three coupled partial differential equations for determining the displacement components v_1, v_2, and v_3.

Method B. In the equations of compatibility (1.104) the strain and bending measures are expressed in terms of the effective contact quantities with the help of the constitutive equations (i.e. (1.135) solved with respect to the strain and bending measures). The resulting three equations are supplemented by the three equations of equilibrium (projection) described in method A (from which the Q_α were eliminated), and with the effective contact forces and couples expressed in terms of the effective contact quantities. We thus obtain a system of six coupled partial differential equations for the determination of the six effective contact quantities.

The resulting equations are in both cases linear in the unknown functions.

Properties of the theory. The present version of the theory of shells was developed by Koiter[22] and Sanders.[13] The properties of the theory may be summarized in the following manner:

(a) The deformations of the shell are described by six deformation measures (three strain measures and three bending measures) which vanish in any displacement as a rigid body of the shell.

(b) The internal forces and moments are described by the six effective contact quantities, and the associated effective contact forces and couples satisfy the six equations of equilibrium (1.111).

(c) The principle of virtual work assumes the simple form (1.127). By means of the principle of virtual work and the constitutive equations (1.135) and with suitable edge conditions, it can be proved that the solution is unique. It can also be proved that Betti's theorem and the theorems of minimum potential energy and minimum complementary energy are valid in the present theory of shells.

(d) A static–geometric analogy can be established.

(e) In the case of axisymmetric bending of shells of revolution, the equations of the present theory reduce to the generally accepted classical equations governing this problem.

If we wish to compare the present theory with other theories of shells, we may ascertain whether these theories possess the properties (a) to (e), these properties being regarded as criteria for the consistency and applicability of the theory in question. The first three criteria are particularly important, and it is essential that they be satisfied if the theory is to be regarded as a consistent one.

It is found that almost all of the earlier theories of shells do not satisfy one or more of the essential criteria (a) to (c). So far as the more recent theories of shells are concerned, it is found that most of these satisfy the essential criteria (a) to (c), but that at least one of the two remaining criteria (d) and (e) are not satisfied.

Finally, it should be noted that the Koiter–Sanders theory possesses an additional advantage, namely that the governing equations can be written in tensor form. In this respect, the theory differs from the theory of Novozhilov,[12] which satisfies all the five criteria (a) to (e), but for which the governing equations cannot be written in tensor form.

Problems

1.1 A surface of revolution is given by the equations

$$x = f(\theta_1) \cos \theta_2,$$

$$y = f(\theta_1) \sin \theta_2, \qquad \text{where} \qquad f(\theta_1) > 0,$$

$$z = \theta_1.$$

(1) Determine the base vectors $\mathbf{a}_1$, $\mathbf{a}_2$, $\mathbf{a}_3$ and the Lamé parameters A_1, A_2 of the surface.

(2) Show that the coordinate curves are the lines of curvature of the surface, and find the principal radii of curvature R_1 and R_2.

(3) Verify that the formulae (1.37) are satisfied when the expressions for A_α, R_α, and $\mathbf{e}_i$ belonging to the surface of revolution are substituted in these equations.

(4) Verify that the equations of Gauss and Codazzi (equations (1.46) and (1.43)) are satisfied when the expressions for A_α and R_α belonging to the surface of revolution are substituted in these equations.

1.2 The middle surface of a circular cylindrical shell is given by the equations

$$x = \theta_1,$$

$$y = R \cos \theta_2,$$

$$z = R \sin \theta_2,$$

where R is the radius of the cylinder.

(1) Determine the strain measures $\varepsilon_{\alpha\beta}$ and the bending measures $\kappa_{\alpha\beta}$ of the cylindrical shell, expressed in terms of the displacements.

(2) Show that the strain and bending measures vanish for the following infinitesimal rigid body displacements of the cylindrical shell:

(a) an arbitrary translation,

(b) a rotation about the X-axis,

(c) a rotation about the Y-axis.

1.3 The middle surface of a shell is given by the following equations

$$x = R[\cos \theta_1 + (\theta_1 - \theta_2) \sin \theta_1],$$
$$y = R[\sin \theta_1 - (\theta_1 - \theta_2) \cos \theta_1],$$
$$z = k\theta_2,$$

where $\theta_1 \geqslant \theta_2$, and R and k are positive constants satisfying the relation

$$R^2 + k^2 = 1.$$

(1) Determine the base vectors $\mathbf{a}_1$, $\mathbf{a}_2$, $\mathbf{a}_3$ and the Lamé parameters of the surface.

(2) Show that the coordinate curves are the lines of curvature, find the principal radii of curvature, and prove that the surface is developable.

(3) Determine the strain measures $\varepsilon_{\alpha\beta}$ and the bending measures $\kappa_{\alpha\beta}$ of the shell, expressed in terms of the displacements.

1.4 In the *membrane theory* of shells, the bending moments, the torsional moments, and the transverse shear forces are neglected (i.e. M_{11}, M_{22}, M_{12}, M_{21}, Q_1, and Q_2 are neglected). The only contact forces and couples that remain are therefore N_{11}, N_{22}, and $N_{12} = N_{21}$.

(1) Formulate the equations of equilibrium according to the membrane theory for a cylindrical shell for which the straight generators of the cylindrical middle surface are parallel to the X-axis, and for which the generating curve is situated in the YZ-plane.

(2) Consider a tube with a circular cross-section (radius R) and length L, and suppose that the tube is completely filled with liquid. Determine the membrane contact forces in the tube for this load. It is assumed that the axis of the tube is horizontal, that the normal component of the contact force is zero at the ends of the tube, and that the shear force is zero in the vertical cross-section at the middle of the tube.

1.5 The middle surface of a conical shell is given by the equations

$$x = \theta_1 \cos \theta_2 \sin \beta,$$
$$y = \theta_1 \sin \theta_2 \sin \beta,$$
$$z = \theta_1 \cos \beta,$$

where $0 < \beta < \pi/2$. It is assumed that the Z-axis is directed vertically downwards, that the shell has a free edge at the parallel circle $z = a$ (where $a > 0$), and that the shell is supported along a lower parallel circle.

(1) Determine the equations of equilibrium of the conical shell according to the membrane theory by specialization of the general equilibrium equations (1.111).

(2) Let the applied surface load be given by

$$p_1 = p_2 = 0, \qquad p_3 = p_0 \cos \theta_2 \cos \beta,$$

where p_0 is a constant. Find the corresponding membrane contact forces in the shell.

1.6 A plane, rectangular, elastic plate of constant thickness h and with the edges given by $x = \pm a/2$ and $y = \pm b/2$ is subjected to the following displacements

$$v_1 = v_2 = 0, \qquad v_3 = -kxy,$$

where v_1, v_2, and v_3 are the components of the displacement in the X-, Y- and Z-directions, respectively, and k is a positive constant.

(1) Determine the effective contact forces and couples in the plate due to the above displacements, and show by substitution in the equations of equilibrium that the distributed surface load corresponding to these effective contact forces and couples is zero.

(2) Show that the effective contact forces and couples derived in (1) are in equilibrium with each of the following two systems of edge loads (i.e. that the equations (1.140) and (1.143) are satisfied on the edge curve in both cases).

(a) $$\mathrm{M}_{21} = 2M \quad \text{for} \quad y = \pm b/2,$$

where

$$M = \frac{Eh^3}{12(1+\nu)}k,$$

and all the other components of the applied edge loads are zero.

(b) In this case, the values of Q_2 and M_{21} on the edges $y = \pm b/2$ are shown in Fig. 1A. All the other components of the applied edge loads are zero.

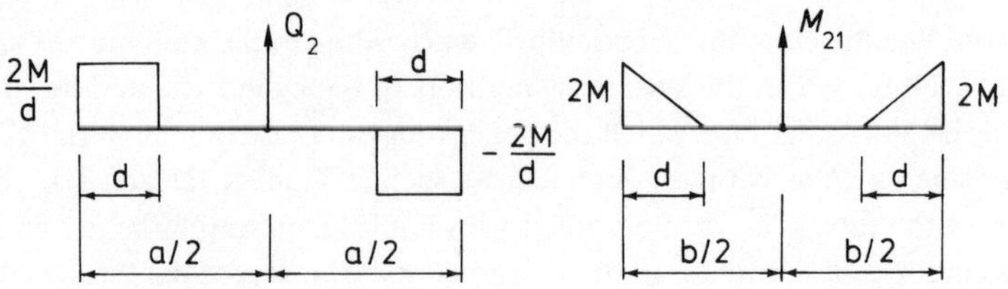

Fig. 1A

(3) Determine the values of the function Φ (the internal reaction) on the boundary curve for both systems of edge loads.

1.7 A plane, rectangular, elastic plate of constant thickness h is simply supported along the sides $x = \pm a/2$ and clamped along the sides $y = \pm b/2$.

(1) By specializing the general equations of the theory of shells, derive the governing differential equation for the normal displacement v_3 corresponding to bending of the plate due to transverse load.

(2) Let the load be given by

$$p_1 = p_2 = 0 \qquad \text{(}X\text{- and }Y\text{-components of load),}$$

$$p_3 = p_0 \cos\left(\frac{m\pi}{a}x\right) \qquad \text{(normal component of load),}$$

where p_0 is a constant, and m is an odd, positive integer. Determine the corresponding displacement v_3 and the bending moments M_{11} and M_{22}.

2

SHELLS OF REVOLUTION

2.1 Introduction

In the preceding chapter we derived a set of general governing equations for shells. If we wish to analyse a particular type of shell structure, the governing equations must be solved analytically or numerically. The *analytical solution* of these complicated differential equations usually encounters considerable difficulties. However, if the geometry of the middle surface is of a sufficiently simple nature, it is sometimes possible to determine exact analytical solutions. This is found to be the case for one of the classical problems in the theory of shells, namely the calculation of shells of revolution. At the beginning of the present century a number of exact analytical solutions had already been derived for this type of shell (H. Reissner 1912, E. Meissner 1913). The problem is of considerable practical interest because of the numerous technical applications of shells of revolution within structural, mechanical, and aerospace engineering.

2.2 Governing equations

We shall derive the governing equations for shells of revolution by a specialization of the general formulae of sections 1.4 to 1.7.

The geometry of the middle surface. The middle surface is a surface of revolution which is formed by rotating a plane curve (the *meridian*) about a straight line in the plane of the curve (the *axis of rotation*) (see Fig. 19). We introduce an orthogonal, Cartesian coordinate system X, Y, Z with the axis of rotation as Z-axis. It will be assumed in the following that the axis of rotation is vertical. The curvilinear coordinates are taken to be

$$\theta_1 = \varphi, \qquad \theta_2 = \theta,$$

where φ is the angle between the axis of rotation and the normal to the meridian, and θ is the angle from the XZ-plane to a plane through the generic point P and the Z-axis (see Fig. 19). The θ_1-curves are therefore the *meridians*, and the θ_2-curves are the *parallel circles*. It is shown in the theory of surfaces that these curves are the *lines of curvature* of the surface of revolution.

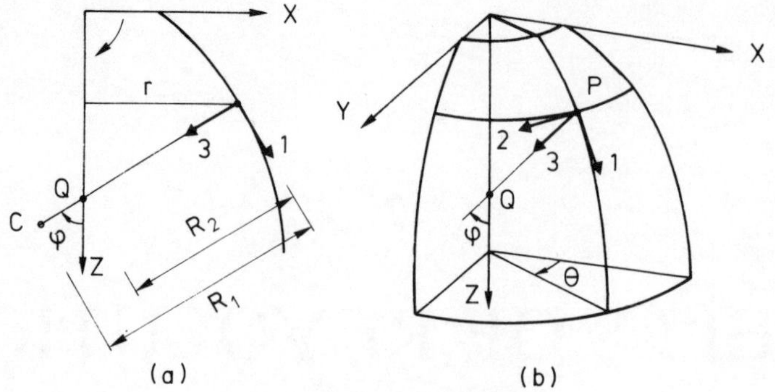

Fig. 19. Surface of revolution

Consider the XZ-plane and the associated meridian in the half-plane $x \geqslant 0$. The positive sense of rotation in the plane is defined in such a way that the angle from the positive X-axis to the positive Z-axis is $+\pi/2$ (see Fig. 19a). It is assumed that the meridian is smooth and has no points of inflection, so that the centres of curvature are all situated on the same side of the curve. The positive direction of the normal to the meridian is chosen as the direction from the point P on the curve to the associated centre of curvature C. The angle φ is measured from the positive Z-axis to the positive direction of the normal, using the positive sense of rotation in the plane. It follows from our previous assumption (no points of inflection) that the arc length s_1 of the meridian is a monotonic function of φ, and it is therefore permissible to use φ as the parameter of the curve. The tangent to the parallel circle at P is perpendicular to the normal to the meridian, the latter line is therefore also a surface normal, and it will be seen that the direction of the unit normal vector to the surface $\mathbf{a}_3$ coincides with the previously defined positive direction of the normal.

It follows from the rotational symmetry of the surface that the principal curvatures $1/R_1$ and $1/R_2$ depend only on φ. It is shown in the theory of surfaces that R_1 equals the distance from P to the corresponding centre of curvature C, while R_2 equals the distance from P to the point of intersection Q between the normal and the axis of rotation. The radii of curvature are taken to be positive if the directions of the corresponding line segments PC and PQ are the same as the positive direction of the normal. It follows from our rule for the determination of the orientation of the normal that R_1 is always positive (cf. the sign convention for the normal curvatures), while R_2 is positive or negative according to whether the concave or the convex side of the meridian faces the axis of rotation.

In order to determine the Lamé parameters, we form the expressions for the line elements along the coordinate curves and use (1.22). Thus

$$\mathrm{d}s_1 = R_1\,\mathrm{d}\varphi = A_1\,\mathrm{d}\varphi, \qquad \mathrm{d}s_2 = A_2\,\mathrm{d}\theta = r\,\mathrm{d}\theta,$$

where r is the radius of the parallel circle (see Fig. 19a). Hence

$$A_1 = R_1(\varphi),\qquad A_2 = r(\varphi) = R_2(\varphi)\sin\varphi, \tag{2.1}$$

in which the last transformation may be verified with the help of Fig. 19a. We note that these expressions for A_1 and A_2 are always positive. We have already shown that R_1 (and therefore A_1) is positive. So far as A_2 is concerned, it is easily verified that the product $R_2 \sin\varphi$ is positive whether the concave or the convex side of the meridian faces the axis of rotation.

From the Codazzi equation $(1.43)_2$ we have

$$\frac{A_{2,1}}{R_1} = \frac{r'}{R_1} = \left(\frac{A_2}{R_2}\right)' = \left(\frac{R_2\sin\varphi}{R_2}\right)' = \cos\varphi, \tag{2.2}$$

or

$$r' = R_1\cos\varphi, \tag{2.3}$$

where

$$(\ )' = \frac{\mathrm{d}}{\mathrm{d}\varphi}(\ ).$$

By means of (2.1) and (2.2) we further obtain

$$\frac{A_{2,1}}{A_1A_2} = \frac{\cot\varphi}{R_2}. \tag{2.4}$$

Inserting the expression $(2.1)_2$ for r in (2.3) and performing the differentiation, we get

$$r' = R'_2\sin\varphi + R_2\cos\varphi = R_1\cos\varphi,$$

or

$$R'_2 = (R_1 - R_2)\cot\varphi. \tag{2.5}$$

It will be assumed in the following that *the thickness of the shell h is constant*. We shall restrict our attention to the *axisymmetric case*, so that the loading, the contact forces and couples, and the displacements are assumed to be functions of φ only (complete rotational symmetry).

Strain and bending measures. It follows from the assumed rotational symmetry that all derivatives with respect to θ $(=\theta_2)$ vanish, and that

$$v_2 = 0,\qquad \varepsilon_{12} = 0,\qquad \kappa_{12} = 0. \tag{2.6}$$

This means that only two displacement components (namely v_1 and v_3) enter into the present problem, and for these we shall here introduce the special notation:

$$v_1 = u,\qquad v_3 = w.$$

Substituting (2.1) and (2.4) into (1.66) and (1.96) we then find for the remaining strain and bending measures

$$\varepsilon_{11} = \frac{1}{R_1}(u' - w), \qquad \varepsilon_{22} = \frac{1}{R_2}(u \cot \varphi - w), \tag{2.7}$$

$$\kappa_{11} = -\frac{1}{R_1}\left(\frac{w' + u}{R_1}\right)', \qquad \kappa_{22} = -\frac{\cot \varphi}{R_1 R_2}(w' + u). \tag{2.8}$$

Equations of equilibrium. Because of the assumed rotational symmetry, the following relations are valid for the complete contact forces and couples and for the applied surface load:

$$\mathrm{N}_{12} = \mathrm{N}_{21} = \mathrm{M}_{12} = \mathrm{M}_{21} = \mathrm{Q}_2 = 0, \qquad p_2 = 0. \tag{2.9}$$

Comparing (2.9) with (1.119) and (1.121), we see that the effective contact forces and couples in the present case are equal to the complete contact forces and couples. It will also be seen that three of the equations of equilibrium (1.111) are satisfied identically (namely the second, fifth and sixth equations). Using (2.1), (2.3) and (2.9), we find for the remaining equations of equilibrium:

$$\begin{aligned} (rN_{11})' - R_1 \cos \varphi \, N_{22} - rQ_1 + rR_1 \mathrm{p}_1 &= 0, \\ rN_{11} + R_1 \sin \varphi \, N_{22} + (rQ_1)' + rR_1 p_3 &= 0, \\ (rM_{11})' - R_1 \cos \varphi \, M_{22} - rR_1 Q_1 &= 0. \end{aligned} \tag{2.10}$$

Constitutive equations. The connection between the contact forces and couples and the strain and bending measures is given by (1.135), i.e.

$$\begin{aligned} N_{11} &= \frac{Eh}{1 - \nu^2}(\varepsilon_{11} + \nu\varepsilon_{22}), \qquad & N_{22} &= \frac{Eh}{1 - \nu^2}(\varepsilon_{22} + \nu\varepsilon_{11}), \\ M_{11} &= D(\kappa_{11} + \nu\kappa_{22}), & M_{22} &= D(\kappa_{22} + \nu\kappa_{11}), \end{aligned} \tag{2.11}$$

where

$$D = \frac{1}{12}\frac{Eh^3}{(1 - \nu^2)}. \tag{2.11a}$$

2.2.1 *Derivation of governing differential equations*

It follows from the linear nature of the governing equations that the general solution of the given inhomogeneous equations can be written as the sum of a *particular integral* (i.e. a solution of the governing equations with the given applied loads p_1, p_3) and the *general solution of the homogeneous equations* (i.e. the governing equations with $p_1 = p_3 = 0$). As a particular integral, it is usually sufficiently accurate to use the solution obtained by means of the *membrane theory* (see section 2.2.3).

We begin with the derivation of the *solution of the homogeneous equations* (the so-called *edge effect*). If we consider a segment of the shell bounded by two parallel circles, this solution corresponds to loading cases in which the applied loads consist solely of forces and moments acting along the two parallel circles.

It will be convenient to introduce the following quantities:

$$\chi = (1/R_1)(w' + u), \qquad (2.12)$$
$$\psi = -R_2 Q_1.$$

χ is the rotation of the tangent to the meridian caused by the displacement (note that $\chi = -\omega_1$, where ω_1 is given by $(1.83)_1$). Substituting (2.12) in (2.8) we get

$$\kappa_{11} = -\chi'/R_1, \qquad \kappa_{22} = -\frac{\cot\varphi}{R_2}\chi. \qquad (2.13)$$

We shall now derive a system of two coupled differential equations for the determination of χ and ψ. In order to obtain the first equation, we consider the equation of moment equilibrium $(2.10)_3$. With the help of (2.3) the first term of this equation can be written in the form

$$(rM_{11})' = rM'_{11} + R_1 \cos\varphi \, M_{11}.$$

Substituting in $(2.10)_3$ we obtain

$$rM'_{11} + R_1 \cos\varphi \, (M_{11} - M_{22}) = R_1 r Q_1.$$

If we divide this equation by $R_1 \sin\varphi$ and use $(2.1)_2$ and $(2.12)_2$, we get

$$\frac{R_2}{R_1} M'_{11} + \cot\varphi \, (M_{11} - M_{22}) = -\psi. \qquad (2.14)$$

This equation can be written in matrix form as follows:

$$\begin{bmatrix} \left(\frac{R_2}{R_1}\partial + \cot\varphi\right) & -\cot\varphi \end{bmatrix} \begin{bmatrix} M_{11} \\ M_{22} \end{bmatrix} = -\psi, \qquad (2.14a)$$

where the differential operator ∂ is determined by

$$\partial(\ ) = \frac{d}{d\varphi}(\ ) = (\ )'.$$

Substituting (2.13) in the last two of equations (2.11), we get

$$\begin{bmatrix} M_{11} \\ M_{22} \end{bmatrix} = D \begin{bmatrix} 1 & \nu \\ \nu & 1 \end{bmatrix} \begin{bmatrix} \kappa_{11} \\ \kappa_{22} \end{bmatrix} = D \begin{bmatrix} 1 & \nu \\ \nu & 1 \end{bmatrix} \begin{bmatrix} -\frac{1}{R_1}\partial \\ -\frac{\cot\varphi}{R_2} \end{bmatrix} \chi.$$

When this expression is inserted in (2.14a), we obtain

$$\begin{bmatrix} \left(\frac{R_2}{R_1}\partial + \cot\varphi\right) & -\cot\varphi \end{bmatrix} \begin{bmatrix} 1 & \nu \\ \nu & 1 \end{bmatrix} \begin{bmatrix} \frac{1}{R_1}\partial \\ \frac{\cot\varphi}{R_2} \end{bmatrix} \chi = \frac{\psi}{D}. \qquad (2.15)$$

Evaluating the left-hand side we have

$$\left[\left(\frac{R_2}{R_1}\partial + \cot\varphi\right) \quad -\cot\varphi\right]\begin{bmatrix}\frac{1}{R_1}\partial + \nu\frac{\cot\varphi}{R_2}\\ \frac{\cot\varphi}{R_2} + \frac{\nu}{R_1}\partial\end{bmatrix}\chi = \frac{\psi}{D},$$

or

$$\begin{aligned}&\frac{R_2}{R_1}\left(\frac{\chi'}{R_1}\right)' + \frac{\cot\varphi}{R_1}\chi' - \frac{\cot^2\varphi}{R_2}\chi\\ &\qquad + \nu\left\{\frac{R_2}{R_1}\left(\frac{\cot\varphi}{R_2}\chi\right)' + \frac{\cot^2\varphi}{R_2}\chi - \frac{\cot\varphi}{R_1}\chi'\right\} = \frac{\psi}{D}.\end{aligned} \tag{2.15a}$$

where

$$\frac{R_2}{R_1}\partial\frac{1}{R_1}\partial\chi \qquad \text{denotes} \qquad \frac{R_2}{R_1}\frac{\mathrm{d}}{\mathrm{d}\varphi}\left(\frac{1}{R_1}\frac{\mathrm{d}\chi}{\mathrm{d}\varphi}\right)$$

(the differential operator acts on the following factors). By using (2.5) we find for the first term in the braces

$$\begin{aligned}\frac{R_2}{R_1}\left(\frac{\cot\varphi}{R_2}\chi\right)' &= \frac{\cot\varphi}{R_1}\chi' - \frac{\chi}{R_1\sin^2\varphi} - \frac{1}{R_1R_2}\cot^2\varphi\,(R_1 - R_2)\chi\\ &= \frac{\cot\varphi}{R_1}\chi' - \frac{\cot^2\varphi}{R_2}\chi - \frac{\chi}{R_1}.\end{aligned}$$

Inserting this expression in (2.15a), we finally obtain

$$L[\chi] - \frac{\nu}{R_1}\chi = \frac{\psi}{D}, \tag{2.16}$$

where, according to (2.15a), the differential operator $L[\quad]$ is given by

$$L[\chi] = \frac{R_2}{R_1}\left(\frac{\chi'}{R_1}\right)' + \frac{\cot\varphi}{R_1}\chi' - \frac{\cot^2\varphi}{R_2}\chi. \tag{2.17}$$

This is one differential equation for χ and ψ.

In order to find a second equation, we begin by deriving a compatibility condition. Multiplying $(2.7)_1$ by R_1 and $(2.7)_2$ by R_2, and subtracting one of the resulting equations from the other, we obtain

$$u' - u\cot\varphi = R_1\varepsilon_{11} - R_2\varepsilon_{22}. \tag{2.18a}$$

Multiplying $(2.7)_2$ by R_2 and differentiating the result, we obtain with the help of (2.5)

$$\begin{aligned}u'\cot\varphi - \frac{u}{\sin^2\varphi} - w' &= (R_2\varepsilon_{22})'\\ &= R_2\varepsilon'_{22} + (R_1 - R_2)\cot\varphi\,\varepsilon_{22}.\end{aligned} \tag{2.18b}$$

We now multiply (2.18a) by cot φ and subtract the result from (2.18b) to obtain the equation

$$R_2\varepsilon_{22}' + R_1 \cot\varphi\,(\varepsilon_{22} - \varepsilon_{11}) = -(w' + u) = -R_1\chi,$$

where $(2.12)_1$ has been used. Dividing by R_1 we get

$$\frac{R_2}{R_1}\varepsilon_{22}' + \cot\varphi\,(\varepsilon_{22} - \varepsilon_{11}) = -\chi. \tag{2.19}$$

It can be shown that this equation is identical with the general equation of compatibility $(1.104)_1$, evaluated for the shell of revolution.

By solving the constitutive equations $(2.11)_{1,2}$ with respect to ε_{11} and ε_{22} we get

$$\begin{bmatrix}\varepsilon_{22}\\ \varepsilon_{11}\end{bmatrix} = \frac{1}{Eh}\begin{bmatrix}1 & -\nu\\ -\nu & 1\end{bmatrix}\begin{bmatrix}N_{22}\\ N_{11}\end{bmatrix}. \tag{2.20}$$

We now express N_{11} and N_{22} in terms of ψ. We consider a section through the shell along a parallel circle, and we project all the internal forces acting along this section onto the axis of rotation. Since the applied surface load is zero

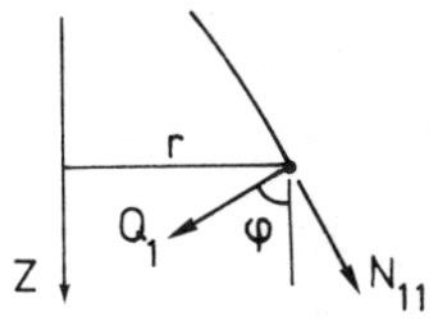

Fig. 20. Components of contact force

($p_1 = p_3 = 0$), this resultant force will be independent of the location of the section, i.e. it will be constant. We therefore have (see Fig. 20)

$$(N_{11}\sin\varphi + Q_1\cos\varphi)2\pi r = P, \tag{2.21}$$

where P denotes the constant resultant force. We now make the further *assumption* that $P = 0$ (a justification of this assumption will be given in section 2.2.3). We therefore obtain

$$N_{11} = -\cot\varphi\, Q_1 = \frac{\cot\varphi}{R_2}\psi, \tag{2.22a}$$

where $(2.12)_2$ has been used. Inserting this expression for N_{11} in the equation of equilibrium $(2.10)_2$, and using $(2.1)_2$ and $(2.12)_2$, we find

$$\begin{aligned} N_{22} &= -\frac{1}{R_1\sin\varphi}[(rQ_1)' + rN_{11}] \\ &= -\frac{1}{R_1\sin\varphi}[(R_2\sin\varphi\; Q_1)' - R_2\cos\varphi\; Q_1] \\ &= -\frac{1}{R_1}(R_2Q_1)', \end{aligned}$$

or

$$N_{22} = \psi'/R_1. \tag{2.22b}$$

Equation (2.19) can be written in the form

$$\begin{bmatrix} \left(\dfrac{R_2}{R_1}\partial + \cot\varphi\right) & -\cot\varphi \end{bmatrix} \begin{bmatrix} \varepsilon_{22} \\ \varepsilon_{11} \end{bmatrix} = -\chi.$$

Inserting (2.20) and (2.22) in this equation, we obtain

$$\begin{bmatrix} \left(\dfrac{R_2}{R_1}\partial + \cot\varphi\right) & -\cot\varphi \end{bmatrix} \begin{bmatrix} 1 & -\nu \\ -\nu & 1 \end{bmatrix} \begin{bmatrix} \dfrac{1}{R_1}\partial \\ \dfrac{\cot\varphi}{R_2} \end{bmatrix} \psi = -Eh\chi. \tag{2.23}$$

It will now be seen that this equation can be obtained from (2.15) if we replace χ and ν by ψ and $-\nu$, respectively, on the left-hand side of (2.15), and replace ψ/D by $-Eh\chi$ on the right-hand side. It follows that (2.23) can be transformed into an equation which is obtained from (2.16) if we make the above replacements in the latter equation, i.e.

$$L[\psi] + \frac{\nu}{R_1}\psi = -Eh\chi. \tag{2.24}$$

This is the second equation for ψ and χ.

As a result of our analysis, we have obtained the equations (2.16) and (2.24), i.e.

$$\begin{aligned} L[\chi] - \frac{\nu}{R_1}\chi &= \frac{\psi}{D}, \\ L[\psi] + \frac{\nu}{R_1}\psi &= -Eh\chi. \end{aligned} \tag{2.25}$$

This is a system of two ordinary, linear differential equations of the second order with variable coefficients for the determination of the functions ψ and χ.

We may eliminate χ by inserting the expression $(2.25)_2$ for χ into $(2.25)_1$. This yields the equation

$$L^2[\psi] + \nu L\left[\frac{\psi}{R_1}\right] - \frac{\nu}{R_1}L[\psi] - \frac{\nu^2}{R_1^2}\psi = -\frac{Eh}{D}\psi, \tag{2.26a}$$

where $L^2[\psi] = L[L[\psi]]$. This is a differential equation of the fourth order for ψ. In a similar manner we may eliminate ψ to obtain the equation

$$L^2[\chi] - \nu L\left[\frac{\chi}{R_1}\right] + \frac{\nu}{R_1}L[\chi] - \frac{\nu^2}{R_1^2}\chi = -\frac{Eh}{D}\chi. \tag{2.26b}$$

We may therefore either solve the system of equations (2.25) or one of the equations (2.26).

In the preceding derivation of the bending theory for shells of revolution, it was assumed that the thickness of the shell is constant. If the thickness is variable, it is possible to derive a system of two differential equations of a type similar to (2.25). In this case, the thickness h and the flexural rigidity D (see (2.11a)) are functions of φ, and the equations assume the form

$$\frac{R_2}{R_1}\left(\frac{\chi'}{R_1}\right)' + \left(\cot\varphi + \frac{R_2}{R_1} 3\frac{h'}{h}\right)\frac{\chi'}{R_1} - \frac{\cot^2\varphi}{R_2}\chi - \frac{\nu}{R_1}\left(1 - \cot\varphi\, 3\frac{h'}{h}\right)\chi = \psi/D,$$

$$\frac{R_2}{R_1}\left(\frac{\psi'}{R_1}\right)' + \left(\cot\varphi - \frac{R_2}{R_1}\frac{h'}{h}\right)\frac{\psi'}{R_1} - \frac{\cot^2\varphi}{R_2}\psi + \frac{\nu}{R_1}\left(1 + \cot\varphi\frac{h'}{h}\right)\psi = -Eh\chi. \tag{2.27}$$

In the following we shall restrict our attention to shells of constant thickness. Some analytical solutions of the equations (2.27) for shells with variable thickness may be found in Flügge's book.[4]

2.2.2 *Calculation of contact forces and couples and of displacements*

Let us assume that we have determined the function ψ by solving equation (2.26a). The contact forces and couples and the displacements can be determined in the following manner. Q_1 is given by $(2.12)_2$, and N_{11} and N_{22} are given by (2.22a and b). χ is determined by (2.24), the bending measures κ_{11} and κ_{22} are determined by (2.13), and the bending moments M_{11} and M_{22} are found from (2.11). In order to formulate the edge conditions, we shall need the horizontal component of the internal force per unit length of the parallel circle. This horizontal component is denoted by $\mathscr{H}$ and is shown in its positive sense in Fig. 21a. From (2.22a) we find that

$$\mathscr{H} = Q_1 \sin\varphi - N_{11}\cos\varphi = Q_1(\sin\varphi + \cos\varphi\cot\varphi),$$

or

$$\mathscr{H} = Q_1/\sin\varphi. \tag{2.28}$$

In order to determine the displacements, it is convenient to calculate the horizontal and vertical components. These components are denoted by $\mathscr{W}$ and $\mathscr{U}$, respectively, and are shown in Fig. 21b in their positive senses. The

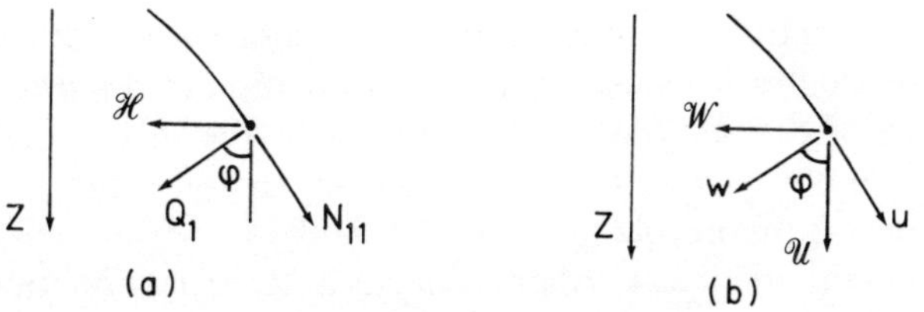

Fig. 21. Horizontal and vertical components of contact forces and displacements

component $\mathscr{W}$ is given by

$$\mathscr{W} = w \sin \varphi - u \cos \varphi = \sin \varphi \, (w - u \cot \varphi),$$

or, from $(2.7)_2$ and (2.20),

$$\mathscr{W} = - \sin \varphi \, R_2 \varepsilon_{22} = - \frac{R_2 \sin \varphi}{Eh} (N_{22} - \nu N_{11}). \tag{2.29}$$

Substituting the expressions (2.22) for N_{11} and N_{22} in (2.29), we get

$$\mathscr{W} = - \frac{\sin \varphi}{Eh} \left(\frac{R_2}{R_1} \psi' - \nu \cot \varphi \, \psi \right). \tag{2.30}$$

The component $\mathscr{U}$ is given by

$$\mathscr{U} = u \sin \varphi + w \cos \varphi .$$

On differentiating this equation, we get

$$\begin{aligned} \mathscr{U}' &= u' \sin \varphi + u \cos \varphi + w' \cos \varphi - w \sin \varphi \\ &= \sin \varphi \, [u' - w + \cot \varphi \, (w' + u)]. \end{aligned}$$

Using (2.7), $(2.12)_1$ and (2.20) we find

$$\mathscr{U}' - R_1 \sin \varphi \, [\varepsilon_{11} + \cot \varphi \, \chi]$$

$$\mathscr{U}' = R_1 \sin \varphi \left(\frac{1}{Eh} (N_{11} - \nu N_{22}) + \cot \varphi \, \chi \right). \tag{2.31}$$

Inserting the expressions (2.22) for N_{11} and N_{22}, we deduce

$$\mathscr{U}' = \sin \varphi \left[\frac{1}{Eh} \left(\frac{R_1}{R_2} \cot \varphi \, \psi - \nu \psi' \right) + R_1 \cot \varphi \, \chi \right]. \tag{2.32}$$

Integrating this equation, we obtain the following formula for $\mathscr{U}$:

$$\mathscr{U}(\varphi) = \mathscr{U}_0 + \int_{\varphi 0}^{\varphi} \mathscr{U}' \, \mathrm{d}\varphi. \tag{2.32a}$$

It will be seen that this expression contains an arbitrary constant $\mathscr{U}_0$ (corresponding to a vertical translation of the shell as a whole).

2.2.3 *Particular integral, membrane theory*

In the previous sections we have studied the solution of the homogeneous equations. We shall now consider the determination of a *particular integral*, i.e. a solution of the governing equations with the given loads p_1 and p_3 (note that it is not required that the particular integral satisfy the prescribed edge conditions of the problem).

It is shown in the theory of ordinary, linear differential equations that if the complete solution of the homogeneous equations is known, then a particular integral can always be determined by means of the method of variation of parameters. However, if the loading and the geometry of the shell are not of a simple nature, the resulting expressions become rather complicated. We shall therefore confine ourselves to the description of a simple approximate method, in which the solution according to the *membrane theory* is used as an approximate particular integral. The membrane solution is easily determined, and it can be used as a sufficiently accurate approximation in most of the cases that arise in practical applications.

It is well known that the membrane theory of shells is based on the assumption that the moments $M_{\alpha\beta}$ and the transverse shear forces Q_α can be neglected. Introducing this assumption in the first two equations of equilibrium (2.10), we obtain

$$\begin{aligned} &(rN^{\mathrm{m}}_{11})' - R_1 \cos\varphi \, N^{\mathrm{m}}_{22} + rR_1 p_1 = 0, \\ &rN^{\mathrm{m}}_{11} + R_1 \sin\varphi \, N^{\mathrm{m}}_{22} + rR_1 p_3 = 0, \end{aligned} \tag{2.33}$$

where the index m indicates that the corresponding quantities are determined by means of membrane theory. Multiplying the first equation (2.33) by $\sin\varphi$ and the second by $\cos\varphi$, and adding the resulting equations, we obtain

$$(\sin\varphi \, rN^{\mathrm{m}}_{11})' + rR_1(p_1 \sin\varphi + p_3 \cos\varphi) = 0,$$

and hence, by integration,

$$N^{\mathrm{m}}_{11}(\varphi) = -\frac{1}{2\pi r \sin\varphi}\left(P + \int_{\varphi 0}^{\varphi} 2\pi r R_1(p_1 \sin t + p_3 \cos t)\,\mathrm{d}t\right). \tag{2.34}$$

This formula for N^{m}_{11} can, alternatively, be derived by expressing the fact that the resultant vertical force on the portion of the shell between the two parallel circles corresponding to the angles φ_0 and φ should vanish (see Fig. 22). It is easily seen that the integral on the right-hand side of (2.34) is the total applied surface load on this part of the middle surface, while the constant P is the resultant vertical force due to the contact forces $N^{\mathrm{m}}_{11}(\varphi_0)$ acting along the parallel circle at φ_0. Having determined N^{m}_{11} by means of (2.34), we find, with the help of $(2.33)_2$, that

$$N^{\mathrm{m}}_{22} = -\left(\frac{R_2}{R_1} N^{\mathrm{m}}_{11} + R_2 p_3\right). \tag{2.35}$$

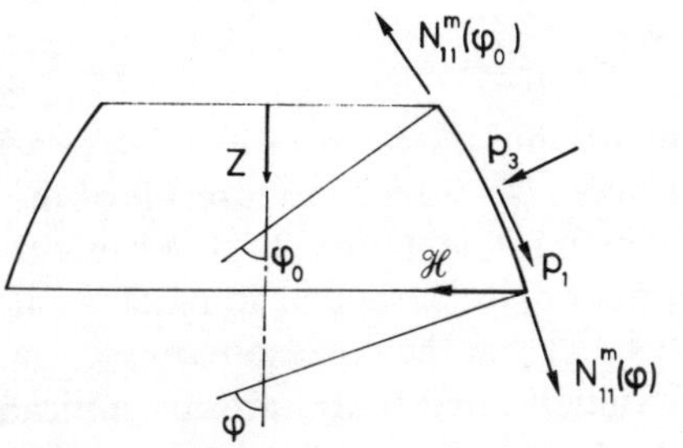

Fig. 22. Membrane contact forces and loads

This completes the determination of the contact forces of the membrane theory, since $N_{12}^{m} = N_{21}^{m} = 0$ because of the rotational symmetry. The horizontal component of the internal force per unit length of parallel circle is given by

$$\mathscr{H}^{m} = -N_{11}^{m} \cos \varphi . \tag{2.36}$$

We now proceed to the determination of the displacements of the membrane theory. Using (2.29) we obtain for the horizontal component $\mathscr{W}^{m}$ of the displacements (see Fig. 21b)

$$\mathscr{W}^{m} = -R_2 \sin \varphi \, \varepsilon_{22} = -\frac{R_2 \sin \varphi}{Eh} (N_{22}^{m} - \nu N_{11}^{m}). \tag{2.37}$$

In order to determine the rotation of the tangent χ^{m}, we use the compatibility condition (2.19) and express the strains in terms of the contact forces by means of (2.20). This gives

$$\chi^{m} = -\frac{R_2}{R_1} \left(\frac{1}{Eh} (N_{22}^{m} - \nu N_{11}^{m}) \right)' + \cot \varphi \, \frac{1+\nu}{Eh} (N_{11}^{m} - N_{22}^{m}). \tag{2.38}$$

Finally, the vertical component $\mathscr{U}^{m}$ is determined by means of (2.31), in which the above expressions for N_{11}^{m}, N_{22}^{m}, and χ^{m} are inserted on the right-hand side.

It follows from the expression (2.34) for N_{11}^{m} that the membrane solution can be regarded as a sum of two contributions. The first of these is due to the vertical component P of the contact forces along the parallel circle at φ_0, while the second is due to the applied surface load. We recall that a contribution of the former type also appeared in connection with the solution of the homogeneous equations (see (2.21) in section 2.2.1). However, the subsequent derivations in that section were based on the assumption that the P-value due to the solution of the homogeneous equations is zero. It will now be seen that it is permissible to introduce this assumption, since the contribution from a vertical force resultant P, as shown in the present section, can be treated in connection with the particular integral (the membrane solution).

Conditions for the use of the membrane solution as a particular integral. We wish to use the membrane solution as an approximate particular integral. In order to assess the accuracy of this approximation, the following considerations may be made. If the rotation χ^{m} due to the membrane solution is substituted in the expressions (2.13) for the bending measures, it will generally be found that the values of these quantities differ from zero. This means that certain bending moments and transverse shear forces arise, in addition to the membrane contact forces N_{11}^{m} and N_{22}^{m}, as a result of the displacements calculated by means of the membrane theory. In order that the membrane solution may be used as an approximate particular integral, we must now demand that these moments be so small that the corresponding normal stresses in the shell (regarded as a three-dimensional body) are insignificant compared with the normal stresses due to the membrane contact forces. It can be shown that this implies that the following conditions must be imposed on the geometry of the shell and on the applied loading (see Hildebrand[7]).

The shell should be thin, and the meridian should be a smooth curve. The principal radii of curvature R_1 and R_2, the thickness of the shell h, and the components p_1 and p_3 of the applied surface load should be continuous functions of φ and should satisfy certain smoothness conditions (this implies that discontinuous changes of slope of the meridian or of thickness of the shell are not allowed). Moreover, the above geometrical and statical quantities should not vary too rapidly along the meridian. The latter condition will be satisfied if the derivative with respect to φ of each of the geometrical quantities R_1, R_2, and h is not large compared with the quantity itself, and if the components p_1 and p_3 of the applied load do not change appreciably over a distance $(R_2h)^{1/2}$ along the meridian (see Hildebrand[7]). If the meridian is a straight line (cf. the case of a conical shell or cylindrical shell), the derivative with respect to φ in the above condition should be replaced by the derivative $L\,\mathrm{d}(\;\;)/\mathrm{d}s_1$, where s_1 is the arc length of the meridian, and L is a characteristic length of the middle surface (e.g. a radius of a parallel circle).

2.2.4 Edge conditions

Let us consider a shell of revolution bounded by two parallel circles and under the action of a certain axisymmetric applied load. The complete solution of the inhomogeneous equations is determined as the sum of a particular integral and the complete solution of the homogeneous equations. Now the particular integral does not generally satisfy the prescribed edge conditions. The solution of the homogeneous equations should therefore be determined in such a way that the sum of this solution and the particular integral satisfies the edge conditions.

The differential equation (2.26a) is of the fourth order, and the solution of the homogeneous equation therefore contains four arbitrary constants. In addition to these we have the constant P in equation (2.34) and the constant $\mathscr{U}_0$ in equation (2.32a) (the vertical resultant of the contact forces along a parallel circle, and the vertical translation of the shell, respectively). The total number of arbitrary constants is therefore six, and these constants are determined by means of six prescribed edge conditions. For example, in the case of purely geometrical edge conditions, one would prescribe the two displacement components and the rotation of the tangent at each of the two boundary curves. Statical edge conditions are expressed by means of the reduced edge loads (see section 1.7). It follows from (1.141) that the first, third and fourth component of the reduced edge loads in the present axisymmetric case are equal to N_{11}, Q_1, and M_{11}, respectively, while the second component vanishes.

The normal case. In most of the cases that arise in practical applications, the vertical resultant of the contact forces along one of the parallel circles is known (the resultant of the contact forces along any other parallel circle can then be determined from equilibrium considerations) and the vertical displacement of one of the boundary curves is also known. This means that the values of the constants P and $\mathscr{U}_0$ are known, so that only four constants remain to be determined (this situation will be called the *normal case*).

In the case of *discontinuities* in the geometrical and/or statical quantities, the above method of solution must be modified, since it was assumed in the derivation of both the solution of the homogeneous equations and the membrane solution that certain conditions of continuity and smoothness were satisfied (the discontinuities in question may be sudden changes of slope of the meridian or of thickness of the shell, sudden changes of applied load, or a ring beam along a parallel circle). In such cases the following procedure may be used. The shell is regarded as being divided into a number of *segments* by means of sections along those parallel circles at which discontinuities occur. The above method of solution can then be used for each of the segments, so that four nontrivial arbitrary constants appear in the normal case in connection with each segment of the shell. It will be shown in the following that four conditions of transition may be formulated at each section, and to these should be added two edge conditions at the upper edge and two at the lower edge. The total number of conditions is therefore equal to the total number of arbitrary constants.

We finally discuss various types of edge conditions that occur frequently in the applications. In this connection it will be assumed that we are dealing with the above *normal case*.

(a) *Clamped edge*. In this case the displacement and the rotation are zero, i.e.

$$\mathscr{W} = 0, \qquad \chi = 0 \tag{2.39}$$

at the edge.

(b) *Simply supported edge*. In this case the displacement and the bending moment are zero, i.e.

$$\mathscr{W} = 0, \qquad M_{11} = 0 \tag{2.40}$$

at the edge.

(c) *Free edge*. At a free edge the forces and the bending moment are zero, i.e.

$$\mathscr{H} = 0, \qquad M_{11} = 0. \tag{2.41}$$

The vertical component of the contact force at the edge also equals zero. The contribution to this quantity from the solution of the homogeneous equations is always zero (see section 2.2.1). The condition concerning the vertical component can therefore be satisfied by an appropriate choice of the constant P in the membrane solution (see (2.34)).

(d) *Ring beam*. We shall confine ourselves to an approximate treatment of the conditions of transition associated with a ring beam. Assuming that the ring beam is slender, we shall neglect the bending stiffness and regard the beam as a perfectly flexible member (a treatment of this problem that includes the effect of bending stiffness has been given by Brøndum-Nielsen,[1] p. 62).

We consider a slender ring beam along a parallel circle. The parallel circle

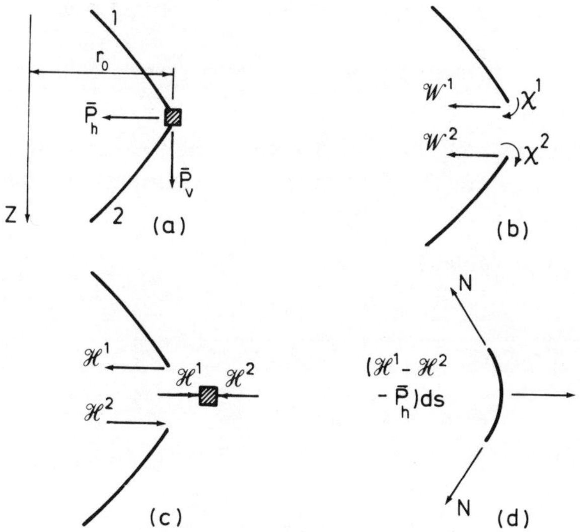

Fig. 23. Conditions at ring beam

divides the shell into two segments, an upper segment 1, and a lower segment 2 (see Fig. 23). Discontinuities such as sudden changes of thickness may occur at the parallel circle. The radius of the parallel circle is denoted by r_0, and it is assumed that the centroids of the cross-sections of the beam coincide with the parallel circle. In the following we shall neglect the cross-sectional dimensions of the beam and regard the area as being concentrated at the centroid. The assumed perfect flexibility of the beam implies that the moments and the shear forces in the beam are zero. The only internal force in the ring beam is therefore the direct force which will be denoted by N.

The applied loads comprise distributed surface loads acting on the two segments. Moreover, it will be assumed that an applied line load acts on the ring beam. This line load is given by the horizontal and vertical components $\bar{P}_h$ and $\bar{P}_v$, measured per unit length of the parallel circle. Fig. 23c shows the internal forces acting between the shell segments and the ring beam. The upper indices 1 and 2 indicate that the corresponding quantities belong to segments 1 and 2, respectively. It will be seen that the total horizontal force per unit length of the parallel circle is given by $\mathscr{H}^1 - \mathscr{H}^2 - \bar{P}_h$ (positive when the force is directed away from the axis of rotation). Projection onto a horizontal line in the direction of the normal to the parallel circle for a small element of the ring beam gives the equation

$$N = r_0(\mathscr{H}^1 - \mathscr{H}^2 - \bar{P}_h). \tag{2.42}$$

We note that the strain of the ring beam equals the strain ε_{22} of the shell segments along the corresponding parallel circle, and the displacement of the ring beam equals the displacement of the shell segments at the parallel circle.

Assuming a linearly elastic beam, we therefore obtain (see 2.29)

$$N = E_b A \varepsilon_{22} = -\frac{E_b A}{r_0} \mathscr{W}^1, \tag{2.43}$$

where E_b is Young's modulus and A is the cross-sectional area of the beam. We can now formulate the following four conditions of transition at the parallel circle:

Continuity of displacements

$$\mathscr{W}^1 = \mathscr{W}^2,$$

Continuity of rotations

$$\chi^1 = \chi^2,$$

Force equilibrium (2.44)

$$\bar{P}_h + \mathscr{H}^2 - \mathscr{H}^1 = \frac{E_b A}{r_0^2} \mathscr{W}^1,$$

Moment equilibrium

$$M_{11}^1 = M_{11}^2.$$

The third equation $(2.44)_3$ is obtained with the help of (2.42) and (2.43). This condition ensures that the resultant horizontal force on the beam element is zero. When the membrane contact forces are determined in such a manner that they are in equilibrium with the applied loads (see section 2.2.3), the resultant vertical force on the ring beam will also be zero. This follows from the fact that the vertical component of the contact forces due to the solution of the homogeneous equations is zero.

2.2.5 *The special case R_1 = constant*

This important case comprises spherical, conical, and cylindrical shells. The second and third terms in each of the two equations (2.26) now cancel each other, and the equations assume the form

$$\begin{aligned} L^2[\psi] - \frac{\nu^2}{R_1^2}\psi &= -\frac{Eh}{D}\psi, \\ L^2[\chi] - \frac{\nu^2}{R_1^2}\chi &= -\frac{Eh}{D}\chi. \end{aligned} \tag{2.45}$$

Noting that

$$\frac{Eh}{D} = Eh \bigg/ \left(\frac{1}{12}\,\frac{Eh^3}{(1-\nu^2)}\right) = 12\,\frac{(1-\nu^2)}{h^2},$$

and introducing the positive constant μ defined by

$$\mu^4 = 12\,\frac{(1-\nu^2)}{h^2} - \frac{\nu^2}{R_1^2}, \tag{2.46}$$

we can write equation $(2.45)_1$ in the form

$$L^2[\psi] + \mu^4\psi = 0. \tag{2.47}$$

Now it is easily verified that this equation can be rewritten in the form

$$L[L[\psi] + \mathrm{i}\mu^2\psi] - \mathrm{i}\mu^2(L[\psi] + \mathrm{i}\mu^2\psi) = 0, \tag{2.48a}$$

or in the form

$$L[L[\psi] - \mathrm{i}\mu^2\psi] + \mathrm{i}\mu^2(L[\psi] - \mathrm{i}\mu^2\psi) = 0, \tag{2.48b}$$

where i denotes the imaginary unit. It follows that the solutions of the two second-order differential equations

$$L[\psi] + \mathrm{i}\mu^2\psi = 0, \qquad L[\psi] - \mathrm{i}\mu^2\psi = 0, \tag{2.49}$$

will also satisfy (2.48a) and (2.48b), respectively, and therefore the original equation (2.47). The solutions of the equations (2.49) are complex functions of the real variable φ. If we form the complex conjugate to $(2.49)_1$, we obtain (since the coefficients of the differential operator L are real)

$$L[\bar{\psi}] - \mathrm{i}\mu^2\bar{\psi} = 0.$$

It follows that, for any solution ψ of $(2.49)_1$, the complex conjugate $\bar{\psi}$ will be a solution of $(2.49)_2$. Now the complete solution of $(2.49)_1$ has the form

$$\psi(\varphi) = Af_1(\varphi) + Bf_2(\varphi), \tag{2.50a}$$

where f_1 and f_2 are the two linearly independent solutions, and A and B are arbitrary complex constants. The complete solution of $(2.49)_2$ can therefore be written in the form

$$\psi(\varphi) = A_1\bar{f}_1(\varphi) + B_1\bar{f}_2(\varphi), \tag{2.50b}$$

since the linear independence of f_1 and f_2 implies that $\bar{f}_1$ and $\bar{f}_2$ will be linearly independent. An arbitrary linear combination of the four functions f_1, f_2, $\bar{f}_1$, $\bar{f}_2$ is therefore a solution of the original equation (2.47), i.e. the real and imaginary parts of the functions f_1 and f_2 constitute four real solutions of this equation. It can be shown that these four real functions are linearly independent, so that they determine the complete solution of (2.47). In order to find the complete solution of the fourth-order differential equation (2.47), it is therefore sufficient to determine the complete (complex) solution of one of the second-order equations (2.49).

2.3 Spherical shell

We shall now derive exact and approximate solutions of the governing equations for some important types of shells of revolution. We first consider a spherical shell of constant thickness. In this case we have the relations

$$R_1 = R_2 = R = \text{constant},$$

and the differential operator $L[\ \ ]$ assumes the form (see (2.17))

$$L[\ \ ] = \frac{1}{R}[(\ \ )'' + \cot\varphi(\ \ )' - \cot^2\varphi(\ \ )]. \tag{2.51}$$

If we introduce the positive constant κ defined by

$$\kappa^4 = \frac{R^2}{4}\mu^4 = 3(1-\nu^2)\frac{R^2}{h^2} - \frac{\nu^2}{4} \tag{2.52}$$

and use the fact that $\psi = -RQ_1$ (see $(2.12)_2$), the first of equations (2.49) assumes the form

$$Q_1'' + Q_1'\cot\varphi - Q_1\cot^2\varphi + 2i\kappa^2 Q_1 = 0. \tag{2.53}$$

As previously explained, the complete solution of this equation also determines the complete solution of the corresponding fourth-order equation (2.47).

If we introduce new variables defined by

$$x = \sin^2\varphi, \qquad F = Q_1/\sin\varphi,$$

it can be shown that (2.53) is transformed into the following differential equation

$$x(x-1)\frac{\mathrm{d}^2F}{\mathrm{d}x^2} + (\tfrac{5}{2}x - 2)\frac{\mathrm{d}F}{\mathrm{d}x} + \frac{1-2i\kappa^2}{4}F = 0. \tag{2.54}$$

This equation is a special case of the so-called hypergeometric differential equation, which has the general form

$$x(1-x)y'' + [\gamma - (\alpha+\beta+1)x]y' - \alpha\beta y = 0, \tag{2.55}$$

where α, β, and γ are constants. A comparison between the two equations (2.54) and (2.55) shows that if we substitute the values

$$\alpha = \frac{3 + (5 + 8i\kappa^2)^{1/2}}{4}, \qquad \beta = \frac{3 - (5 + 8i\kappa^2)^{1/2}}{4}, \qquad \gamma = 2,$$

into (2.55), then the resulting equation will become identical with (2.54).

The determination of the hypergeometric functions is a classical problem in mathematical analysis, and the properties of these functions (series expansions, asymptotic behaviour, etc.) are well known (see Copson[40]). Exact analytical solutions for the spherical shell can therefore be determined by means of the hypergeometric functions. However, this method of solution requires considerable computation effort. No tables are available for the general hypergeometric functions (because of the dependence of these functions on the three parameters α, β, and γ), and it is therefore necessary to calculate the values of the functions by means of appropriate series expansions (power series). Now in the case of thin shells, it is found that these series converge slowly, so that a large number of terms of the series must be computed in order to obtain results of a sufficient accuracy. This method, therefore, requires the use of an electronic computer. We shall not enter into a detailed description of the method (such a description may be found in the books of Flügge[4] and Timoshenko and Woinowsky-Krieger[15]), but we shall instead consider two approximate methods for spherical shells which simplify the calculation considerably.

Before we leave the exact solution for the spherical shell, it should be mentioned that the solution may be represented in an alternative form with the help of Legendre functions (see Seide,[14] section 6.4).

Approximate solution, type 1 (Geckeler's method). We consider an edge zone of a thin spherical shell, i.e. a region of the shell in the vicinity of a boundary curve (parallel circle). It will be assumed that the shell is not a shallow one in this region, i.e. that the angle φ is not a very small quantity in the edge zone. If such a shell is analysed by means of the above exact method, it will be found that the contact forces and couples and the displacements have the character of damped oscillations which decay rapidly with increasing distance from the edge. From the mathematical point of view, this behaviour is characterized by the fact that the first derivative with respect to φ of each of the above quantities is large compared with the quantity itself, the second derivative is large compared with the first derivative, etc. In the differential operator $L[\quad]$ (see (2.51)) we shall therefore omit terms containing the function and the first derivative and retain the term containing the second derivative. (In order that this approximation should be permissible, it must be assumed that cot φ is not a large quantity, i.e. that φ is not a small angle. This assumption will be satisfied if the shell is not shallow.) Thus we obtain the approximate expression

$$L[\quad] \simeq \frac{1}{R}(\quad)''.$$

With the help of this approximation, the differential equations (2.47) and (2.53) assume the form

$$Q_1'''' + 4\kappa^4 Q_1 = 0, \tag{2.56a}$$

$$Q_1'' + 2\mathrm{i}\kappa^2 Q_1 = 0. \tag{2.56b}$$

These are now equations with constant coefficients, and the solution is therefore easily determined. On substituting the trial solution

$$Q_1(\varphi) = C\,\mathrm{e}^{r\varphi}, \tag{2.57}$$

into (2.56b), we find the auxiliary equation

$$r^2 + 2\mathrm{i}\kappa^2 = 0, \tag{2.58}$$

the roots of which are given by

$$r = \pm(1 - \mathrm{i})\kappa. \tag{2.59}$$

If we form a linear combination of the real and imaginary parts of the two solutions corresponding to (2.57) we obtain, as previously explained, the complete solution of the fourth-order equation (2.56a), i.e.

$$Q_1(\varphi) = e^{-\kappa\varphi}(C_1 \cos \kappa\varphi + C_2 \sin \kappa\varphi) + e^{\kappa\varphi}(C_3 \cos \kappa\varphi + C_4 \sin \kappa\varphi), \quad (2.60)$$

where $C_1, \ldots, C_4$ are arbitrary constants. It will be seen that the four solutions on the right-hand side of (2.60) have the form of damped oscillations. Now, for a thin shell, we have the relation $R/h \gg 1$. We shall therefore omit the last term on the right-hand side of the formula (2.52) for κ^4, i.e. we shall use the approximate expression

$$\kappa = [3(1 - \nu^2)]^{1/4}(R/h)^{1/2}, \quad (2.61)$$

and we see that $\kappa^2 \gg 1$ for a thin shell. It follows that the first two terms on the right-hand side of (2.60) are rapidly damped out when φ increases, while the last two terms are damped out when φ decreases.

It will be shown in section 2.6 that the present approximate method can be used not only for spherical shells but also for arbitrary thin shells of revolution that are not too shallow. The further development of the method (determination of the remaining statical and geometrical quantities, etc.) will therefore be given in that section.

Approximate solution, type 2 (shallow shells). The above approximate method for a spherical shell cannot be used in a neighbourhood of the point $\varphi = 0$ (the apex of the shell), where the angle φ is very small. In this neighbourhood the quantity cot φ will be large, and the second and third terms in the differential operator $L[\ \ ]$ can no longer be neglected (see (2.51)).

This neighbourhood of the apex may be regarded as a shallow shell, and another type of approximation is therefore introduced which applies to shallow shells. This approximate method will be treated in section 3.5 in connection with the theory of shallow shells.

2.4 Conical shell

We shall now consider a conical shell (Fig. 24) of constant thickness. The following quantities are used as curvilinear coordinates

$$\theta_1 = s, \qquad \theta_2 = \theta,$$

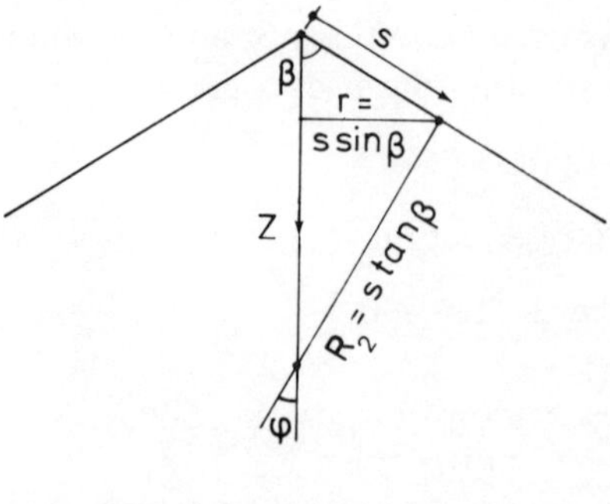

Fig. 24. Conical shell

where s denotes the distance from the apex measured along the straight generator (the meridian), and the meaning of the angle θ is indicated in Fig. 19b. From the relations

$$ds_1 = ds = d\theta_1, \qquad ds_2 = r\,d\theta = A_2\,d\theta,$$

the following expressions are obtained for the Lamé parameters and the principal radii of curvature (see (1.22) and Fig. 19a):

$$\begin{aligned} A_1 &= 1, & A_2 &= r = s\sin\beta, \\ 1/R_1 &= 0, & R_2 &= s\tan\beta. \end{aligned} \tag{2.62}$$

The equations for the conical shell can be obtained by introducing the geometrical quantities (2.62) into the equations of the general theory of shells in Chapter 1. However, we can derive these equations in a simpler way by applying the following limiting process to the equations of the general shell of revolution in section 2.2. We assume, first, that R_1 is constant (i.e. independent of φ); we then introduce the arc length s_1 of the meridian as a new independent variable; and finally we let $R_1 \to \infty$. In this way we arrive at the equations of the conical shell.

By using the limiting process and the ensuing relations

$$\begin{gathered} \frac{1}{R_1}\frac{d}{d\varphi}(\ \ ) = \frac{d}{ds_1}(\ \ ) \to \frac{d}{ds}(\ \ ) = (\ \ )^{\cdot}, \\ \varphi \to \pi/2 - \beta, \qquad \cot\varphi \to \tan\beta, \qquad R_2 \to s\tan\beta \end{gathered} \tag{2.63}$$

where a dot denotes differentiation with respect to s, we obtain the following equations for the conical shell.

Strain and bending measures. From (2.7) and (2.8) we find

$$\begin{aligned} \varepsilon_{11} &= u^{\cdot}, & \varepsilon_{22} &= (1/s)(u - w\cot\beta), \\ \kappa_{11} &= -w^{\cdot\cdot}, & \kappa_{22} &= -(1/s)w^{\cdot}. \end{aligned} \tag{2.64}$$

Equations of equilibrium. From (2.10) we obtain

$$\begin{aligned} &(sN_{11})^{\cdot} - N_{22} + sp_1 = 0, \\ &N_{22}\cot\beta + (sQ_1)^{\cdot} + sp_3 = 0, \\ &(sM_{11})^{\cdot} - M_{22} - sQ_1 = 0. \end{aligned} \tag{2.65}$$

The constitutive equations are still given by (2.11).

We now consider the *solution of the homogeneous equations*. The expressions (2.12) and (2.13) take the form

$$\begin{aligned} \chi &= w^{\cdot}, & \psi &= -sQ_1\tan\beta, \\ \kappa_{11} &= -\chi^{\cdot}, & \kappa_{22} &= -(1/s)\chi. \end{aligned} \tag{2.66}$$

If we apply the limiting process to (2.17), we get

$$L[\ \ ] = \tan\beta\,[s(\ \ )^{\cdot\cdot} + (\ \ )^{\cdot} - (1/s)(\ \ )]. \tag{2.67}$$

Introducing the differential operator

$$L_c[\ ] = s(\ )^{\cdot\cdot} + (\ )^{\cdot} - (1/s)(\ ), \tag{2.68}$$

we find that the second equation (2.49) can be written in the form

$$\tan\beta L_c[-sQ_1 \tan\beta] + i\mu^2 sQ_1 \tan\beta = 0. \tag{2.69}$$

We now introduce the positive constant ρ defined by

$$\rho^2 = \frac{[12(1-\nu^2)]^{1/2}}{h} \cot\beta. \tag{2.70}$$

It follows from (2.46) that

$$\mu^2 = \frac{[12(1-\nu^2)]^{1/2}}{h} = \rho^2 \tan\beta,$$

since $1/R_1 = 0$ for the conical shell. Equation (2.69) can therefore be written in the form

$$L_c[sQ_1] - i\rho^2 sQ_1 = 0. \tag{2.71}$$

If we solve this equation to determine Q_1, the remaining statical and geometrical quantities can then be calculated from the following equations (which are obtained from (2.22), (2.28), $(2.66)_2$, (2.11), (2.24), (2.29), and (2.31), and by appropriate use of the limiting process):

$$\begin{gathered}
N_{11} = -Q_1 \tan\beta, \qquad N_{22} = -\tan\beta (sQ_1)^{\cdot}, \\
\mathcal{H} = Q_1/\cos\beta, \\
M_{11} = -D(\chi^{\cdot} + \nu\chi/s), \qquad M_{22} = -D(\chi/s + \nu\chi^{\cdot}), \\
\chi = \frac{\tan^2\beta}{Eh} L_c[sQ_1], \\
\mathcal{W} = \frac{\sin\beta}{Eh} s \tan\beta[(sQ_1)^{\cdot} - \nu Q_1], \\
\mathcal{U}^{\cdot} = \sin\beta \left(\frac{1}{Eh} [-Q_1 + \nu(sQ_1)^{\cdot}] + \chi \right).
\end{gathered} \tag{2.72}$$

We have seen that the determination of the transverse shear force Q_1 is reduced to the integration of the differential equation (2.71). If we introduce the new independent variable

$$x = 2\rho\sqrt{s}, \tag{2.73}$$

so that

$$\frac{d}{ds}(\ ) = \frac{\rho}{\sqrt{s}} \frac{d}{dx}(\ ), \qquad \frac{d^2}{ds^2}(\ ) = \frac{\rho^2}{s} \frac{d^2}{dx^2}(\ ) - \frac{\rho}{2\sqrt{s^3}} \frac{d}{dx}(\ ),$$

equation (2.71) will assume the form

$$x^2 \frac{d^2G}{dx^2} + x \frac{dG}{dx} - (ix^2 + 4)G = 0,$$

where

$$G = sQ_1. \tag{2.74}$$

This equation is closely related to Bessel's differential equation. If we consider a differential equation of the form

$$x^2 \frac{d^2G}{dx^2} + x \frac{dG}{dx} - (ix^2 + n^2)G = 0, \tag{2.75}$$

where n is a positive integer or zero, the complete solution is written in the form

$$G(x) = A \operatorname{be}_n x + B \operatorname{ke}_n x, \tag{2.76}$$

where the complex functions $\operatorname{be}_n x$ and $\operatorname{ke}_n x$ are called *Kelvin functions* of order n, and A and B are arbitrary complex constants. The following notation is introduced for the real and imaginary parts of these functions:

$$\operatorname{be}_n x = \operatorname{ber}_n x + i \operatorname{bei}_n x, \qquad \operatorname{ke}_n x = \operatorname{ker}_n x + i \operatorname{kei}_n x, \tag{2.77}$$

Note that ber and bei denote the real and imaginary parts of the function be, and similarly for the function ke. The properties of the Kelvin functions are known, and they have been tabulated by several authors (see for example Abramowitz and Stegun,[38] section 9.9).

A comparison between (2.74) and (2.75) now shows that the solution of the equation (2.71) for the conical shell is determined by the Kelvin functions of the second order. It was mentioned in section 2.2.5 that the complete solution of the governing fourth-order differential equation (2.47) was given by a linear combination of the real and imaginary parts of the solutions of the corresponding second-order differential equation $(2.49)_2$ or, in the case of the conical shell, of the real and imaginary parts of the functions $\operatorname{be}_2 x$ and $\operatorname{ke}_2 x$. Such a linear combination can be written in the form

$$\begin{aligned} sQ_1 &= a_1 \operatorname{ker}_2 x + b_1 \operatorname{kei}_2 x + a_2 \operatorname{ber}_2 x + b_2 \operatorname{bei}_2 x \\ &= \operatorname{Re}[(a_1 - ib_1)(\operatorname{ker}_2 x + i \operatorname{kei}_2 x) + (a_2 - ib_2)(\operatorname{ber}_2 x + i \operatorname{bei}_2 x)] \\ &= \operatorname{Re}(A_1 \operatorname{ke}_2 x + A_2 \operatorname{be}_2 x), \end{aligned} \tag{2.78}$$

where

$$A_1 = a_1 - ib_1, \qquad A_2 = a_2 - ib_2$$

are arbitrary complex constants. The contributions to the remaining statical and geometrical quantities from this solution can be found by inserting (2.78) in the equations (2.72).

It can be proved by means of certain recurrence relations that the Kelvin functions of order n and their derivatives can be expressed in terms of the Kelvin functions of order zero and their first derivatives, i.e. in terms of the four functions

$$\text{be}\,x, \qquad \text{ke}\,x, \qquad \text{be}'\,x, \qquad \text{ke}'\,x.$$

(Note that the functions of order zero are usually written without lower indices, and that primes in connection with the Kelvin functions denote derivatives with respect to the argument x.) For example, the connection between the functions of the second order and those of order zero is given by the formulae

$$\begin{aligned} g_2(x) &= -\left(g_0(x) + \mathrm{i}\frac{2}{x}g_0'(x)\right), \\ g_2'(x) &= \frac{2}{x}g_0(x) - \left(1 - \mathrm{i}\frac{4}{x^2}\right)g_0'(x), \end{aligned} \tag{2.79}$$

where g_0, g_2 denote either the functions be x, $\text{be}_2\,x$ or the functions ke x, $\text{ke}_2\,x$.

If $f(x)$ denotes one of the functions appearing on the left-hand side of (2.72), it can be shown by means of the recurrence relations that each of these quantities can be written in the form

$$\begin{aligned} f(x) &= \text{Re}[A_1(P_0\,\text{ke}\,x + P_1\,\text{ke}'\,x) + A_2(P_0\,\text{be}\,x + P_1\,\text{be}'\,x)] \\ &= \text{Re}\left\{[P_0 \quad P_1]\begin{bmatrix} \text{ke}\ x & \text{be}\ x \\ \text{ke}'\,x & \text{be}'\,x \end{bmatrix}\begin{bmatrix} A_1 \\ A_2 \end{bmatrix}\right\} \\ &= [R_0 \quad J_0 \quad R_1 \quad J_1]\begin{bmatrix} \text{ker}\ x & \text{kei}\ x & \text{ber}\ x & \text{bei}\ x \\ -\text{kei}\ x & \text{ker}\ x & -\text{bei}\ x & \text{ber}\ x \\ \text{ker}'\,x & \text{kei}'\,x & \text{ber}'\,x & \text{bei}'\ x \\ -\text{kei}'\,x & \text{ker}'\,x & -\text{bei}'\,x & \text{ber}'\ x \end{bmatrix}\begin{bmatrix} a_1 \\ b_1 \\ a_2 \\ b_2 \end{bmatrix}, \end{aligned} \tag{2.80}$$

where the following notation has been introduced for the real and imaginary parts of the complex numbers appearing in these expressions:

$$\begin{aligned} A_1 &= a_1 - \mathrm{i}b_1, & A_2 &= a_2 - \mathrm{i}b_2, \\ P_0 &= R_0 + \mathrm{i}J_0, & P_1 &= R_1 + \mathrm{i}J_1, \\ \text{be}\,x &= \text{ber}\,x + \mathrm{i}\,\text{bei}\,x, & \text{ke}\,x &= \text{ker}\,x + \mathrm{i}\,\text{kei}\,x. \end{aligned} \tag{2.80a}$$

If we insert (2.78) in (2.72) and use (2.79) and the representation (2.80), we find, by means of calculations, the details of which will be omitted, that the contribution to the required statical and geometrical quantities from the *solution of the homogeneous equations* can be written in the following manner:

Assembling the required quantities in the vector

$$\mathbf{f}(s)^{\mathrm{T}} = [\,\mathscr{H} \quad M_{11} \quad \mathscr{W} \quad \chi \quad N_{11} \quad N_{22} \quad M_{22} \quad Q_1], \tag{2.81}$$

we then obtain the matrix equation

$$\mathbf{f}^{\mathrm{h}}(s) = \mathbf{K}(s)\mathbf{B}(x)\mathbf{F}(x)\mathbf{a}, \tag{2.82}$$

where the index h indicates that this is a contribution from the solution of the homogeneous equations. **K** is a diagonal matrix given by

$$\mathbf{K}(s) = \left[\frac{1}{s\cos\beta} \;\; \frac{2}{x^2} \;\; \frac{\sin\beta\tan\beta}{Eh} \;\; \frac{(\tan^2\beta)\rho^2}{Eh} \;\; \frac{\tan\beta}{s} \;\; \frac{\tan\beta}{2s} \;\; \frac{2}{x^2} \;\; \frac{1}{s}\right],$$

B, **F**, and **a** are given by

$$\mathbf{B}(x) = \begin{bmatrix} -1 & 0 & 0 & -2/x \\ 0 & -2(1-\nu) & (4/x)(1-\nu) & x \\ (1+\nu) & 0 & -x/2 & (2/x)(1+\nu) \\ 0 & -1 & 2/x & 0 \\ 1 & 0 & 0 & 2/x \\ -2 & 0 & x & -4/x \\ 0 & 2(1-\nu) & -(4/x)(1-\nu) & \nu x \\ -1 & 0 & 0 & -2/x \end{bmatrix}, \qquad (2.82a)$$

$$\mathbf{F}(x) = \begin{bmatrix} \ker x & \operatorname{kei} x & \operatorname{ber} x & \operatorname{bei} x \\ -\operatorname{kei} x & \ker x & -\operatorname{bei} x & \operatorname{ber} x \\ \ker' x & \operatorname{kei}' x & \operatorname{ber}' x & \operatorname{bei}' x \\ -\operatorname{kei}' x & \ker' x & -\operatorname{bei}' x & \operatorname{ber}' x \end{bmatrix}, \qquad \mathbf{a} = \begin{bmatrix} a_1 \\ b_1 \\ a_2 \\ b_2 \end{bmatrix},$$

and

$$x = 2\rho\sqrt{s}.$$

In order to find the complete solution for a given loading case, it is necessary to add a *particular integral* to the above solution (2.82) of the homogeneous equations. It will be recalled that the membrane solution can generally be used as an approximate particular integral (see section 2.2.3). The resulting expression for the complete solution can then be used to formulate the edge conditions (see section 2.2.4 and the example in section 2.7).

Graphs of the functions ber x, bei x, ker x, and kei x are shown in Fig. 25a. It will be seen that these functions have the character of damped oscillations, the functions ker x and kei x being damped out with increasing x, and the functions ber x and bei x being damped out with decreasing x. It will be seen from the expression (2.82) for the solution of the homogeneous equations and the formulae for **F** and **a** that the constants a_1 and b_1 appear as coefficients of the functions ker x, kei x, and their derivatives, while a_2 and b_2 appear as coefficients of ber x, bei x, and their derivatives. If we now consider a conical shell bounded by two parallel circles (see Fig. 25b), then the terms containing a_1 and b_1 represent an edge effect which originates from the upper edge and for which the displacements and the contact forces and couples are damped out with increasing distance from this edge. Similarly, the terms containing a_2 and b_2 represent an edge effect which originates from the lower edge.

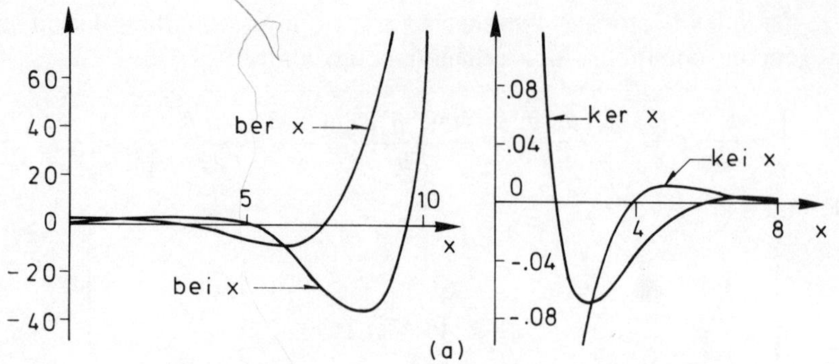

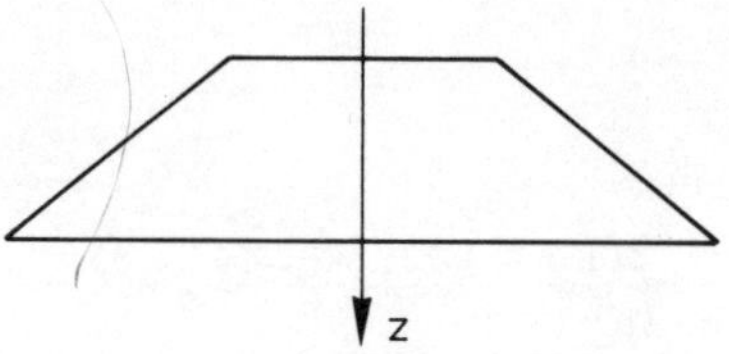

Fig. 25. (a) Kelvin functions and (b) segment of shell

Tables of the Kelvin functions

$$\text{ber } x, \quad \text{bei } x, \quad \text{ker } x, \quad \text{kei } x,$$
$$\text{ber}' x, \quad \text{bei}' x, \quad \text{ker}' x, \quad \text{kei}' x,$$

for values of the argument in the interval $0 < x \leqslant 10$ may be found, for example, in Hütte.[41] For x values greater than 10, asymptotic expressions for these functions may be used (see Abramowitz and Stegun,[38] section 9.9). These expressions may conveniently be calculated by means of a programmable pocket calculator. If, alternatively, the calculation is performed on a large-scale electronic computer, the possibility of using an existing library program for the Kelvin functions should be borne in mind, since such programs are occasionally available at large computer installations.

2.5 Circular cylindrical shell

We shall now consider a circular cylindrical shell of constant thickness. The axisymmetric loading case with which we shall deal occurs in practice, for example, in connection with liquid containers (tanks).

The middle surface is a circular cylinder with radius R, and the Z-axis of the coordinate system coincides with the axis of the cylinder (the axis of rotation). The following quantities are used as curvilinear coordinates:

$$\theta_1 = z/R = \eta, \qquad \theta_2 = \theta \tag{2.83}$$

(see Fig. 26). It will be seen that the coordinate curves are the straight

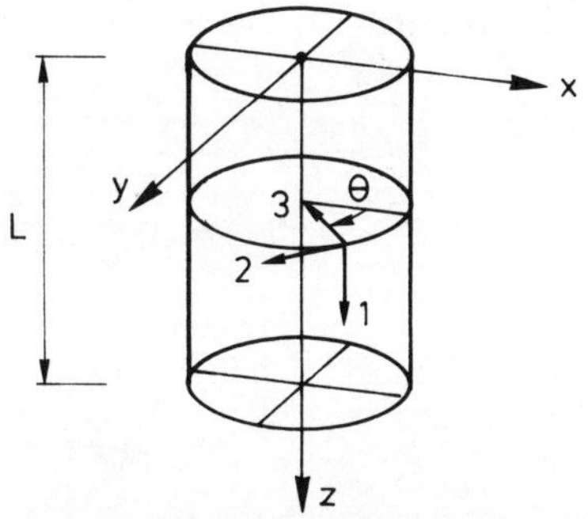

Fig. 26. Cylindrical shell

generators and the parallel circles. The values of the Lamé parameters and the principal radii of curvature are found to be (see section 1.2.1)

$$A_1 = A_2 = R, \qquad \frac{1}{R_1} = 0, \qquad R_2 = R. \tag{2.84}$$

We shall now derive the governing equations by means of a limiting process similar to that used for the conical shell. We obtain the equations of the cylindrical shell from the general equations in section 2.1 by first putting R_1 = constant in the latter equations, then introducing the quantity s_1/R as a new independent variable (where s_1 is the arc length of the meridian), and finally letting $R_1 \to \infty$ and $\varphi \to \pi/2$ (see Fig. 19a), so that

$$\sin\varphi \to 1, \qquad \cos\varphi \to 0, \qquad \cot\varphi \to 0,$$
$$\frac{1}{R_1}\frac{\mathrm{d}}{\mathrm{d}\varphi}(\ ) = \frac{\mathrm{d}}{\mathrm{d}s_1}(\ ) \to \frac{\mathrm{d}}{\mathrm{d}z}(\ ) = \frac{1}{R}\frac{\mathrm{d}}{\mathrm{d}\eta}(\ ) = \frac{1}{R}(\ )^{\cdot}. \tag{2.85}$$

In the present section, dots denote derivatives with respect to η. Using the limiting process, we then find the following equations for the cylindrical shell.

Strain and bending measures. From (2.7) and (2.8) we find

$$\begin{aligned} \varepsilon_{11} &= u^{\cdot}/R, & \varepsilon_{22} &= -w/R, \\ \kappa_{11} &= -w^{\cdot\cdot}/R^2, & \kappa_{22} &= 0. \end{aligned} \tag{2.86}$$

Equations of equilibrium. From (2.10) we get

$$\begin{aligned} N^{\cdot}_{11} + Rp_1 &= 0, & N_{22} + Q^{\cdot}_1 + Rp_3 &= 0, \\ M^{\cdot}_{11} - RQ_1 &= 0. \end{aligned} \tag{2.87}$$

We now turn our attention to the *solution of the homogeneous equations*. Equations (2.12) and (2.13) take the form

$$\begin{aligned} \chi &= w^{\cdot}/R, & \psi &= -RQ_1, \\ \kappa_{11} &= -\chi^{\cdot}/R, & \kappa_{22} &= 0, \end{aligned} \tag{2.88}$$

while (2.17) and (2.47) reduce to

$$L[\ \ ] = \frac{1}{R}(\ \ )^{\cdot\cdot}, \tag{2.89}$$

$$Q_1^{\cdot\cdot\cdot\cdot} + 4\kappa^4 Q_1 = 0, \tag{2.90}$$

where

$$4\kappa^4 = R^2\mu^4 = 12(1 - \nu^2)\left(\frac{R}{h}\right)^2. \tag{2.90a}$$

The remaining statical and geometrical quantities expressed in terms of Q_1 are found from (2.22), $(2.11)_{3,4}$, (2.24), and (2.30). Thus

$$\begin{aligned} N_{11} &= 0, & N_{22} &= -Q_1^{\cdot}, \\ M_{11} &= -\frac{D}{EhR} Q_1^{\cdot\cdot\cdot}, & M_{22} &= \nu M_{11}, \\ \chi &= \frac{1}{Eh} Q_1^{\cdot\cdot}, & w &= \frac{R}{Eh} Q_1^{\cdot}. \end{aligned} \tag{2.91}$$

Equation (2.90) is a linear differential equation of the fourth order with constant coefficients, which has the same form as the equation for a beam on an elastic foundation. Substituting a trial solution of the form

$$Q_1 = C\,\mathrm{e}^{r\eta}, \tag{2.92}$$

we find the auxiliary equation

$$r^4 + 4\kappa^4 = 0, \tag{2.93}$$

the roots of which are given by

$$r = \pm(1 \pm \mathrm{i})\kappa. \tag{2.93a}$$

The complete solution therefore has the form

$$Q_1(\eta) = \mathrm{e}^{-\kappa\eta}(a_1 \cos \kappa\eta + b_1 \sin \kappa\eta) + \mathrm{e}^{\kappa\eta}(a_2 \cos \kappa\eta + b_2 \sin \kappa\eta), \tag{2.94}$$

where a_1, b_1, a_2, and b_2 are arbitrary constants. It is easy to verify that (2.94) can be written in the alternative form

$$Q_1(\eta) = \mathrm{Re}(A_1\,\mathrm{e}^{r_1\eta} + A_2\,\mathrm{e}^{-r_1\eta}), \tag{2.95}$$

where

$$A_1 = a_1 - \mathrm{i}b_1, \qquad A_2 = a_2 + \mathrm{i}b_2, \tag{2.95a}$$

are arbitrary complex constants, and

$$r_1 = (-1 + \mathrm{i})\kappa, \tag{2.95b}$$

$$\mathrm{e}^{\pm r_1\eta} = \mathrm{e}^{\mp\kappa\eta}(\cos \kappa\eta \pm \mathrm{i} \sin \kappa\eta). \tag{2.95c}$$

The independent variable η is proportional to the distance from the upper edge (see (2.83) and Fig. 26). In the second term on the right-hand side of (2.95), we shall introduce a new independent variable, which is proportional to the distance from the lower edge, i.e.

$$\zeta = \frac{1}{R}(L - z) = \frac{L}{R} - \eta, \tag{2.96}$$

where L is the length of the cylinder. We then find

$$A_2\,e^{-r_1\eta} = A_2\,e^{-r_1(L/R-\zeta)} = (A_2\,e^{-r_1L/R})e^{r_1\zeta}. \tag{2.97}$$

The term in parentheses in the last expression may be regarded as a modified arbitrary constant. The solution (2.95) can therefore be written in the form

$$Q_1(\eta) = \mathrm{Re}(A_1\,e^{r_1\eta} + A_2\,e^{r_1\zeta}), \tag{2.98}$$

where A_2 now denotes the modified arbitrary constant. If we insert (2.98) in the equations (2.91) and use the relation

$$(e^{r_1\zeta})^{\cdot} = \frac{d}{d\eta}(e^{r_1(L/R-\eta)}) = -r_1e^{r_1\zeta}, \tag{2.99}$$

we find the following contributions to the remaining statical and geometrical quantities from the solution of the homogeneous equations:

$$\begin{aligned} N_{22} &= -\mathrm{Re}(A_1r_1e^{r_1\eta} - A_2r_1e^{r_1\zeta}) = -\frac{Eh}{R}w, \\ M_{11} &= -\frac{D}{EhR}\mathrm{Re}(A_1r_1^3e^{r_1\eta} - A_2r_1^3e^{r_1\zeta}), \\ \chi &= \frac{1}{Eh}\mathrm{Re}(A_1r_1^2e^{r_1\eta} + A_2r_1^2e^{r_1\zeta}), \end{aligned} \tag{2.100}$$

where

$$r_1^2 = -2\kappa^2 i, \qquad r_1^3 = 2(1 + i)\kappa^3,$$

(see (2.95b)). The statical and geometrical quantities which are required for the formulation of the edge conditions are now assembled in the vector

$$\mathbf{f}(\eta)^{\mathrm{T}} = [Q_1 \quad M_{11} \quad w \quad \chi]. \tag{2.101}$$

By using a representation similar to (2.80), we then obtain the following expression for **f** corresponding to the *complete solution*:

$$\mathbf{f}(\eta) = \mathbf{f}^{\mathrm{p}}(\eta) + \mathbf{KA}\begin{bmatrix} f(\eta) & g(\eta) & 0 & 0 \\ -g(\eta) & f(\eta) & 0 & 0 \\ 0 & 0 & f(\zeta) & g(\zeta) \\ 0 & 0 & -g(\zeta) & f(\zeta) \end{bmatrix}\begin{bmatrix} a_1 \\ b_1 \\ a_2 \\ b_2 \end{bmatrix}. \tag{2.102}$$

In this equation $\mathbf{f}^{\mathrm{p}}$ denotes the contribution from the particular integral, while the last term is the contribution from the solution of the homogeneous equations. The matrices **K** and **A** and the functions f and g are given by

$$\mathbf{K} = \begin{bmatrix} 1 & 0 & 0 & 0 \\ 0 & R/(2\kappa) & 0 & 0 \\ 0 & 0 & \dfrac{R\kappa}{Eh} & 0 \\ 0 & 0 & 0 & \dfrac{2\kappa^2}{Eh} \end{bmatrix}, \qquad \mathbf{A} = \begin{bmatrix} 1 & 0 & 1 & 0 \\ -1 & -1 & 1 & 1 \\ -1 & 1 & 1 & -1 \\ 0 & -1 & 0 & -1 \end{bmatrix}, \tag{2.102a}$$

$$f(\eta) = \mathrm{e}^{-\kappa\eta} \cos \kappa\eta, \qquad g(\eta) = \mathrm{e}^{-\kappa\eta} \sin \kappa\eta.$$

We shall now introduce *dimensionless quantities* (indicated by the tilde sign $\sim$ written above the relevant symbol). Let p_0 be a constant, characteristic loading intensity (of dimension force per unit area), and let us put

$$Q_1 = \frac{p_0 R}{\kappa} \tilde{Q}_1, \qquad M_{11} = \frac{p_0 R^2}{2\kappa^2} \tilde{M}_{11}, \qquad w = \frac{p_0 R^2}{Eh} \tilde{w},$$

$$\chi = 2\kappa \frac{p_0 R}{Eh} \tilde{\chi}, \qquad N_{11} = p_0 R \tilde{N}_{11}, \qquad N_{22} = p_0 R \tilde{N}_{22}, \tag{2.103}$$

$$M_{22} = \frac{p_0 R^2}{2\kappa^2} \tilde{M}_{22}.$$

The equations (2.102) can then be written in the simple dimensionless form

$$\begin{bmatrix} \tilde{Q}_1 \\ \tilde{M}_{11} \\ \tilde{w} \\ \tilde{\chi} \end{bmatrix} = \begin{bmatrix} \tilde{Q}_1^{\mathrm{p}} \\ \tilde{M}_{11}^{\mathrm{p}} \\ \tilde{w}^{\mathrm{p}} \\ \tilde{\chi}^{\mathrm{p}} \end{bmatrix} + \begin{bmatrix} f_\eta & g_\eta & f_\zeta & g_\zeta \\ -(f_\eta - g_\eta) & -(f_\eta + g_\eta) & (f_\zeta - g_\zeta) & (f_\zeta + g_\zeta) \\ -(f_\eta + g_\eta) & (f_\eta - g_\eta) & (f_\zeta + g_\zeta) & -(f_\zeta - g_\zeta) \\ g_\eta & -f_\eta & g_\zeta & -f_\zeta \end{bmatrix} \begin{bmatrix} \tilde{a}_1 \\ \tilde{b}_1 \\ \tilde{a}_2 \\ \tilde{b}_2 \end{bmatrix}, \tag{2.104}$$

where $f_\eta = f(\eta)$, $f_\zeta = f(\zeta)$, etc., and the dimensionless arbitrary constants are given by

$$a_\alpha = \frac{p_0 R}{\kappa} \tilde{a}_\alpha, \qquad b_\alpha = \frac{p_0 R}{\kappa} \tilde{b}_\alpha, \qquad \alpha = 1, 2.$$

We also have

$$\begin{aligned} \tilde{N}_{22} &= \tilde{N}_{22}^{\mathrm{p}} - (\tilde{w} - \tilde{w}^{\mathrm{p}}), \\ \tilde{N}_{11} &= \tilde{N}_{11}^{\mathrm{p}}, \qquad \tilde{M}_{22} = \nu \tilde{M}_{11}. \end{aligned} \tag{2.104a}$$

In order to determine the arbitrary constants, four *edge conditions* (two at each edge) must be formulated, whereby a system of four linear equations with unknowns $\tilde{a}_1$, $\tilde{b}_1$, $\tilde{a}_2$, and $\tilde{b}_2$ is obtained. It will be seen that the coefficients of $\tilde{a}_1$ and $\tilde{b}_1$ in (2.104) depend only on the functions $f(\eta)$ and $g(\eta)$. These coefficients are therefore damped out with increasing η (see (2.102a)) and correspond to an edge effect originating from the upper edge. Similarly, the terms containing $\tilde{a}_2$ and $\tilde{b}_2$ are damped out with increasing ζ and correspond to an edge effect originating from the lower edge. If the damping is so large that the influence of the edge effect from one edge is negligible at the other edge (this condition will be satisfied for $L/\sqrt{(Rh)} \geqslant 3.5$), then the edge effect from the opposite edge may, to a good approximation, be omitted in the formulation of the edge conditions. In this case the terms containing $\tilde{a}_2$ and $\tilde{b}_2$ may be omitted from the equations expressing the edge conditions at the upper edge ($\eta = 0$), while the terms containing $\tilde{a}_1$ and $\tilde{b}_1$ may be omitted from the equations associated with the lower edge ($\zeta \doteq 0$). The equations for the determination of the arbitrary constants are thus separated into two independent systems, each of which consists of two linear equations with two unknowns.

Particular integral. It will be assumed that the conditions of section 2.2.3 are satisfied so that the *membrane solution* can be used as an approximate particular integral. In section 2.2.3 we studied the membrane solution for an arbitrary shell of revolution. However, because of the simplicity of the equations of the cylindrical shell, a direct derivation of the required membrane solution will be preferred in the present case.

Since the relations $Q_1 = M_{11} = M_{22} = 0$ are valid in the membrane theory, we obtain from the equations of equilibrium (2.87):

$$N_{11}^{\mathrm{m}} = N_{11}^{\mathrm{m}}(0) - R\int_0^\eta p_1(\eta)\,\mathrm{d}\eta, \qquad N_{22}^{\mathrm{m}} = -Rp_3. \tag{2.105}$$

Equation (2.37) assumes the form

$$w^{\mathrm{m}} = -R\varepsilon_{22} = -\frac{R}{Eh}(N_{22}^{\mathrm{m}} - \nu N_{11}^{\mathrm{m}}),$$

which, by means of $(2.105)_2$, may be written

$$w^{\mathrm{m}} = \frac{R^2}{Eh}(p_3 + \nu N_{11}^{\mathrm{m}}/R). \tag{2.106}$$

Using this expression for w^{m} in $(2.88)_1$ we get

$$\chi^{\mathrm{m}} = \dot{w}^{\mathrm{m}}/R = \frac{R}{Eh}(\dot{p}_3 + \nu \dot{N}_{11}^{\mathrm{m}}/R). \tag{2.107}$$

We shall now introduce dimensionless quantities defined by (see (2.103))

$$\begin{aligned} p_3(\eta) &= p_0\tilde{p}(\eta),\\ N_{11}^{\mathrm{m}}(\eta) &= p_0R\tilde{n}(\eta), \end{aligned} \tag{2.108}$$

where p_0 is the above characteristic loading intensity. We then find

$$\tilde{\mathbf{f}}^{m}(\eta) = \begin{bmatrix} \tilde{Q}_1^{m} \\ \tilde{M}_{11}^{m} \\ \tilde{w}^{m} \\ \tilde{\chi}^{m} \end{bmatrix} = \begin{bmatrix} 0 \\ 0 \\ \tilde{p}(\eta) + \nu\tilde{n}(\eta) \\ \dfrac{1}{2\kappa}(\tilde{p}^{\cdot} + \nu\tilde{n}^{\cdot}) \end{bmatrix},$$

$$\tilde{N}_{11}^{m} = \tilde{n}(\eta), \qquad \tilde{N}_{22}^{m} = -\tilde{p}(\eta), \qquad \tilde{M}_{22}^{m} = 0. \tag{2.109}$$

The membrane solution obtained can be used as an approximate particular integral in the formulae (2.104) for the complete solution.

2.6 Geckeler's method

We now return to the type 1 approximate method, which was discussed briefly in section 2.3. In the literature, this method is often referred to as Geckeler's method. It was mentioned in section 2.3 that the method could be used not only for spherical shells but also for arbitrary thin shells of revolution. It will be recalled that the method is based on a property of the solution of the homogeneous equations, namely that the derivative of each of the displacements and the contact forces and couples is substantially greater than the quantity itself. We shall therefore simplify the equations by omitting, so far as possible, all terms with these quantities except those containing the highest derivatives.

We consider a thin shell of revolution, for which the meridian satisfies the conditions of section 2.2 but is otherwise arbitrary. In the following derivation, it will be assumed that the thickness of the shell is constant, but it will be explained later that the method can also be used in the case of a moderate variation of the thickness of the shell along the meridian. It is also assumed that the shell is bounded by two boundary curves, namely an upper and a lower parallel circle. In the same way as in section 2.3, it will be assumed that the shell is not too shallow.

We shall now introduce the above simplification of the governing equations. If we perform the differentiation of the part in the parentheses in the first term on the right-hand side of (2.17) by the product rule, and then retain only the term containing the second derivative of the operand, we will obtain the approximate expression

$$L[\quad] \simeq \frac{R_2}{R_1^2}(\quad)''. \tag{2.110a}$$

By using the formula $\psi = -R_2 Q_1$ we find

$$L^2[\psi] \simeq -\frac{R_2^2}{R_1^4}(R_2 Q_1)'''' \simeq -\frac{R_2^3}{R_1^4} Q_1'''', \tag{2.110b}$$

where the last expression is obtained if we perform the differentiations in the term $(R_2Q_1)''''$ according to the product rule and then omit derivatives of lower orders of Q_1. We shall now derive the equation that results from the simplification of (2.26a). Inserting the approximate expressions (2.110a and b) for L and L^2, and retaining only the term with the fourth derivative of Q_1 on the left-hand side of the equation, we get

$$-\frac{R_2^3}{R_1^4} Q_1'''' = \frac{Eh}{D} R_2 Q_1. \tag{2.111}$$

One might object that the term containing Q_1 on the right-hand side of (2.111) should also be omitted in comparison with the term containing the fourth derivative. However, this would be an erroneous argument. In order to arrive at the approximate equation (2.111), we have omitted a number of terms containing Q_1 and its derivatives of order up to and including the third. Among these, the term with Q_1 has the form

$$\left(\frac{\nu^2 R_2}{R_1^2} - \frac{\cot^4 \varphi}{R_2}\right) Q_1.$$

The numerical value of the ratio between this term and the term on the right-hand side of (2.111) does not exceed $k(h/R_{\min})^2$, where

$$k = \frac{1}{6(1-\nu^2)} \max(\nu^2, \cot^4 \varphi), \qquad \text{and} \qquad R_{\min} = \min(R_1, R_2).$$

For a thin shell, we have the relation $(h/R_{\min})^2 \ll 1$. The term with Q_1 that has been retained in the simplified equation (2.111) is therefore substantially greater than the term omitted, and the suggested omission of the term on the right-hand side of the equation would therefore result in significant errors.

It will be seen that the approximate differential equation (2.111) is much simpler than the original equation (2.26a), but (2.111) is still an equation with variable coefficients, since R_1 and R_2 generally vary with φ. We shall now introduce a further simplification so as to obtain an equation with constant coefficients. We shall assume that the quantities R_1 and R_2 vary so slowly along the meridian that they can be taken to be approximately constant in the boundary zone within which the edge effect is significant.

By using the relation

$$\frac{1}{R_1} Q_1' = \frac{1}{R_1} \frac{dQ_1}{d\varphi} = \frac{dQ_1}{ds},$$

where s denotes the arc length of the meridian, we may write equation (2.111) in the following simplified form (which applies to a boundary zone within which R_1 and R_2 are taken to be constant)

$$\frac{d^4 Q_1}{ds^4} + \frac{12(1-\nu^2)}{h^2 R_2^2} Q_1 = 0, \tag{2.112}$$

where the expression (2.11a) for D has been used. The equations expressing the remaining statical and geometrical quantities in terms of Q_1 may be simplified in a similar manner.

We shall now introduce new independent variables η and ζ, which are proportional to the arc lengths measured from the upper and lower edges, respectively. Thus

$$\eta = \frac{1}{R_{2u}}(s - s_u), \qquad \zeta = \frac{1}{R_{2l}}(s_l - s) \tag{2.113}$$

(see Fig. 27), where the indices u and l denote values associated with the upper and lower edges, respectively.

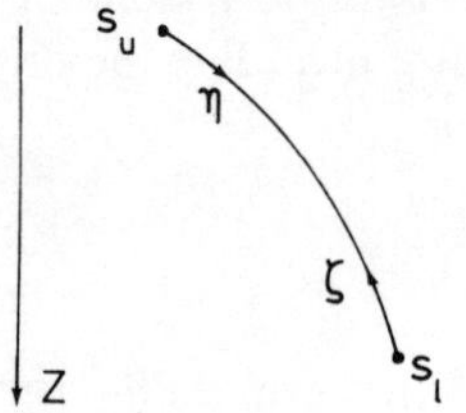

Fig. 27. Meridian section showing variables η and ζ

We first consider the edge effect originating from the *upper edge*. In the differential equation (2.112) we replace R_2 by the constant value R_{2u} at the upper edge, and we also introduce the independent variable η instead of s. Using the relation

$$\frac{d}{ds}(\ ) = \frac{1}{R_{2u}}\frac{d}{d\eta}(\ ),$$

we then obtain the equation

$$\frac{d^4 Q_1}{d\eta^4} + 12(1 - \nu^2)\left(\frac{R_{2u}}{h}\right)^2 Q_1 = 0,$$

or

$$\frac{d^4 Q_1}{d\eta^4} + 4\kappa_u^4 Q_1 = 0, \tag{2.114}$$

in which the constant κ_u is positive and defined by

$$\kappa_u^4 = 3(1 - \nu^2)\left(\frac{R_{2u}}{h}\right)^2. \tag{2.114a}$$

The contribution from this solution to the remaining statical and geometrical quantities is found from the following equations (which are obtained from (2.22), (2.24), (2.30), (2.11), (2.13), and (2.28) by omitting lower-order derivatives and replacing R_2 by R_{2u}):

$$N_{11} = -(\cot\varphi)Q_1, \qquad N_{22} = -\mathrm{d}Q_1/\mathrm{d}\eta,$$
$$\chi = \frac{1}{Eh}\frac{\mathrm{d}^2Q_1}{\mathrm{d}\eta^2}, \qquad \mathscr{W} = (\sin\varphi)\frac{R_{2u}}{Eh}\frac{\mathrm{d}Q_1}{\mathrm{d}\eta},$$
$$M_{11} = -\frac{D}{EhR_{2u}}\frac{\mathrm{d}^3Q_1}{\mathrm{d}\eta^3}, \qquad M_{22} = \nu M_{11}, \tag{2.115}$$
$$\mathscr{H} = Q_1/\sin\varphi.$$

Equation (2.114) is a simple differential equation of the fourth order with constant coefficients. By solving (2.114) and inserting the result in (2.115), we obtain an approximate solution of the homogeneous equations for the arbitrary shell of revolution.

We shall now compare (2.114) and (2.115) with the corresponding equations (2.90) and (2.91) for the circular cylindrical shell. It will be seen that the differential equation for Q_1 and the equations for N_{22}, χ, M_{11}, and M_{22} for the shell of revolution have exactly the same form as the corresponding equations for a circular cylindrical shell with radius $R = R_{2u}$, and with the same values of h, E, and ν as those of the shell of revolution. In the case of the cylindrical shell, we found that the complete solution could be written in matrix form as shown in (2.102). Because of the analogy between the equations of the cylindrical shell and the approximate equations of the shell of revolution, we conclude that a similar matrix representation can be used for the shell of revolution (see below).

The *edge effect* originating from the *lower edge* can be treated in a similar manner. In this case R_2 is replaced by the constant value R_{21} belonging to the lower edge, and equation (2.112) assumes the form

$$\frac{\mathrm{d}^4Q_1}{\mathrm{d}\zeta^4} + 4\kappa_1^4Q_1 = 0, \tag{2.116}$$

where

$$\kappa_1^4 = 3(1-\nu^2)\left(\frac{R_{21}}{h}\right)^2 \tag{2.116a}$$

defines the κ value of the lower edge. Corresponding to (2.115) we here find a set of equations which can be obtained from (2.115) by replacing the index u by l and replacing derivatives with respect to η by derivatives with respect to ζ, and by changing the sign of the equations that contain derivatives of odd orders of Q_1 (since $\mathrm{d}(\ )/\mathrm{d}\eta = -\mathrm{d}(\ )/\mathrm{d}\zeta$).

The *complete approximate solution* for the arbitrary shell of revolution can now be represented in a similar way to the solution (2.102) for the cylindrical shell. It should be noted, however, that the last term of (2.102) (i.e. the contribution from the solution of the homogeneous equations) is replaced, in the case of the shell of revolution, by a sum of two contributions (from the upper and lower edges, respectively), since the quantities R_1, R_2, and κ

generally have different values at the two edges. If we define the vector **f** by

$$\mathbf{f}^{\mathrm{T}}(\eta) = [Q_1 \quad M_{11} \quad \mathscr{W} \quad \chi \quad N_{22}], \tag{2.117}$$

we find the formulae

$$\mathbf{f}(\eta) = \begin{bmatrix} 0 \\ 0 \\ \mathscr{W}^{\mathrm{m}} \\ \chi^{\mathrm{m}} \\ N_{22}^{\mathrm{m}} \end{bmatrix} + \left\lceil \begin{matrix} 1 \\ \dfrac{R_{2\mathrm{u}}}{2\kappa_{\mathrm{u}}} \\ \dfrac{s_\varphi R_{2\mathrm{u}}\kappa_{\mathrm{u}}}{Eh} \\ \dfrac{2\kappa_{\mathrm{u}}^2}{Eh} \\ \kappa_{\mathrm{u}} \end{matrix} \right\rfloor \begin{bmatrix} 1 & 0 \\ -1 & -1 \\ -1 & 1 \\ 0 & -1 \\ 1 & -1 \end{bmatrix} \begin{bmatrix} f_\eta^{\mathrm{u}} & g_\eta^{\mathrm{u}} \\ -g_\eta^{\mathrm{u}} & f_\eta^{\mathrm{u}} \end{bmatrix} \begin{bmatrix} a_1 \\ b_1 \end{bmatrix}$$

$$+ \left\lceil \begin{matrix} 1 \\ \dfrac{R_{2\mathrm{l}}}{2\kappa_{\mathrm{l}}} \\ \dfrac{s_\varphi R_{2\mathrm{l}}\kappa_{\mathrm{l}}}{Eh} \\ \dfrac{2\kappa_{\mathrm{l}}^2}{Eh} \\ \kappa_{\mathrm{l}} \end{matrix} \right\rfloor \begin{bmatrix} 1 & 0 \\ 1 & 1 \\ 1 & -1 \\ 0 & -1 \\ -1 & 1 \end{bmatrix} \begin{bmatrix} f_\zeta^{\mathrm{l}} & g_\zeta^{\mathrm{l}} \\ -g_\zeta^{\mathrm{l}} & f_\zeta^{\mathrm{l}} \end{bmatrix} \begin{bmatrix} a_2 \\ b_2 \end{bmatrix}, \tag{2.118a}$$

$$\begin{aligned} &\mathscr{H} = \mathscr{H}^{\mathrm{m}} + Q_1/\sin\varphi, \\ &N_{11} = N_{11}^{\mathrm{m}} - (\cot\varphi)Q_1, \qquad M_{22} = \nu M_{11}, \end{aligned} \tag{2.118b}$$

where

$$\begin{aligned} &s_\varphi = \sin\varphi, \\ &f_\eta^{\mathrm{u}} = \mathrm{e}^{-\kappa_{\mathrm{u}}\eta}\cos\kappa_{\mathrm{u}}\eta, \qquad g_\eta^{\mathrm{u}} = \mathrm{e}^{-\kappa_{\mathrm{u}}\eta}\sin\kappa_{\mathrm{u}}\eta, \\ &f_\zeta^{\mathrm{l}} = \mathrm{e}^{-\kappa_{\mathrm{l}}\zeta}\cos\kappa_{\mathrm{l}}\eta, \qquad g_\zeta^{\mathrm{l}} = \mathrm{e}^{-\kappa_{\mathrm{l}}\zeta}\sin\kappa_{\mathrm{l}}\zeta, \end{aligned} \tag{2.118c}$$

and the signs ⌈ ⌋ denote a diagonal matrix. It is here assumed that the membrane solution (denoted by an upper index m) is used as an approximate particular integral.

A *moderate variation* of the *thickness of the shell* along the meridian can be accounted for in the following manner. The thickness h is replaced by the value h_{u} belonging to the upper edge in the first diagonal matrix on the right-hand side of (2.118a), and by the value h_{l} belonging to the lower edge in the second diagonal matrix. We also replace h by h_{u} and h_{l}, respectively, in the formulae (2.114a) and (2.116a) for κ_{u} and κ_{l}. Apart from these modifications, the formulae remain unchanged.

The above-mentioned analogy between the approximate equations of the shell of revolution and the equations of a circular cylindrical shell can be interpreted physically as a replacement of the given shell of revolution by an equivalent circular cylindrical shell for the purpose of an approximate calculation of the edge effect. It will be seen that this method represents a far-reaching simplification, in which the effects of the curvature of the meridian, for example, are neglected, and only the most significant features of the behaviour of the shell of revolution are accounted for, namely (1) the bending in the direction of the meridian, and (2) the elastic support due to changes of length of the fibres along the parallel circles. In spite of the many approximations, it will be found that the results according to Geckeler's method agree reasonably well with the exact solution in the case of thin shells (see the following example).

2.7 Numerical example, conical shell

We consider a conical shell of constant thickness and bounded by two parallel circles. The lower edge is simply supported, and a ring beam is attached to the upper edge of the shell. It is assumed that Poisson's ratio is zero ($\nu = 0$). The dimensions of the shell are shown in Fig. 28.

The applied load is vertical and consists of a surface load that is uniformly distributed over the middle surface and a line load acting along the upper edge. We wish to determine the contact forces and couples in the shell due to this load. We shall first determine the solution by means of the exact theory for conical shells derived in section 2.4 and shall then find an approximate solution with the help of Geckeler's method.

The distributed surface load is determined by

$$p_1 = p_0 \cos \beta, \qquad p_3 = p_0 \sin \beta, \tag{2.119}$$

where p_0 is a constant. The line load along the upper edge is given by

$$\bar{P}_v = \tfrac{1}{2} p_0 l_1, \qquad \bar{P}_h = 0. \tag{2.120}$$

Since the conditions for the use of the membrane theory as an approximate particular integral are satisfied (see section 2.2.3), we shall first determine the *membrane solution*. The equations for the conical shell can be obtained from the equations (2.34) to (2.37) for the general shell of revolution by means of

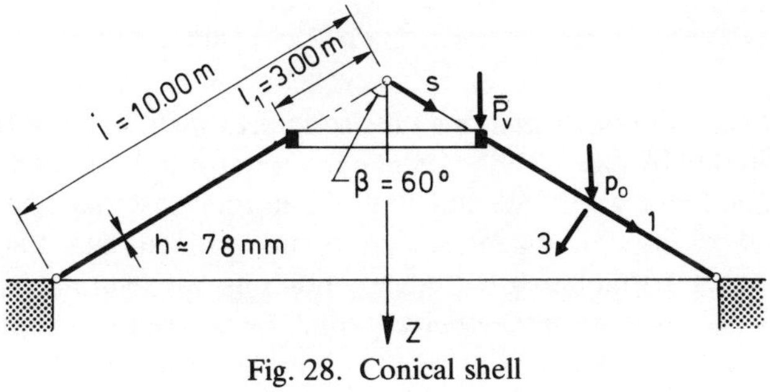

Fig. 28. Conical shell

a limiting process in which we (1) assume that R_1 is constant, (2) introduce the arc length s_1 of the meridian as a new independent variable, and (3) let $R_1 \to \infty$. Using the relations $\mathrm{d}s_1 = R_1\,\mathrm{d}\varphi \to \mathrm{d}s$, $\sin\varphi \to \cos\beta$ and $r \to s\sin\beta$, which are valid as a result of the limiting process, from (2.34) we obtain

$$N_{11}^{\mathrm{m}} = -\frac{1}{2\pi s \sin\beta\cos\beta}\left(P + \int_{l_1}^{s} 2\pi s p_0 \sin\beta\,\mathrm{d}s\right),$$

in which P is the total line load acting along the upper edge, i.e.

$$P = 2\pi l_1 \sin\beta\,\bar{\mathrm{P}}_\upsilon = \pi l_1^2 \sin\beta\, p_0.$$

Using this value of P we find

$$N_{11}^{\mathrm{m}} = -\frac{1}{2}\,\frac{p_0 s}{\cos\beta}. \tag{2.121a}$$

Since $1/R_1 = 0$ for the conical shell, equation (2.35) assumes the form

$$N_{22}^{\mathrm{m}} = -p_0 s \tan\beta\sin\beta, \tag{2.121b}$$

and (2.36) and (2.37) give

$$\mathscr{H}^{\mathrm{m}} = \tfrac{1}{2} p_0 s \tan\beta, \tag{2.121c}$$

$$\mathscr{W}^{\mathrm{m}} = \frac{p_0}{Eh}\, s^2 \tan\beta\sin^2\beta. \tag{2.121d}$$

It is now convenient to introduce *dimensionless quantities* defined by

$$\upsilon = s/l,$$

$$\begin{aligned} \tilde{N}_{11} &= N_{11}/(p_0 l), & \tilde{N}_{22} &= N_{22}/(p_0 l),\\ \tilde{\mathscr{H}} &= \mathscr{H}/(p_0 l), & \tilde{Q}_1 &= Q_1/(p_0 l),\\ \tilde{M}_{11} &= M_{11}/(p_0 l^2), & \tilde{\mathscr{W}} &= \mathscr{W} Eh/(p_0 l^2). \end{aligned} \tag{2.122}$$

The membrane solution can then be written in the following dimensionless form

$$\begin{aligned} \tilde{N}_{11}^{\mathrm{m}} &= -\frac{1}{2}\,\frac{\upsilon}{\cos\beta}, & \tilde{N}_{22}^{\mathrm{m}} &= -\upsilon\sin\beta\tan\beta,\\ \tilde{\mathscr{H}} &= \tfrac{1}{2}\upsilon\tan\beta, & \tilde{\mathscr{W}}^{\mathrm{m}} &= \upsilon^2\sin^2\beta\tan\beta. \end{aligned} \tag{2.123}$$

(a) *Solution of the exact equations for the conical shell*. We shall first determine the solution by means of the exact equations for the conical shell derived in section 2.4. Since we are dealing with the normal case (see section 2.2.4), the solution of the homogeneous equations will contain four arbitrary constants. When formulating the edge conditions, *we shall neglect the influence of the edge effect from the opposite edge*. Because of this approximation, certain errors are introduced but, as shown in the following, these errors will be small in the present problem.

Using the numerical values shown in Fig. 28, we find

$$l/h = 128, \qquad \cot\beta = 1/\sqrt{3},$$

and from (2.70) and (2.73),

$$\rho^2 = \frac{\sqrt{12}}{h}\frac{1}{\sqrt{3}} = \frac{2}{h},$$

$$x = 2\rho\sqrt{s} = 2\sqrt{(2/h)}\sqrt{(l\upsilon)} = 32\sqrt{\upsilon}. \tag{2.124}$$

Edge conditions. The shell is simply supported along the *lower edge*. We therefore have the edge conditions (see (2.40))

$$\mathscr{W} = 0, \qquad M_{11} = 0. \tag{2.125}$$

We shall now neglect the terms with a_1 and b_1 in the representation (2.82) of the solution of the homogeneous equations, as these terms represent contributions from the edge effect originating from the upper edge. Using the dimensionless quantities (2.122) and dimensionless arbitrary constants defined by

$$[\tilde{a}_1 \quad \tilde{b}_1 \quad \tilde{a}_2 \quad \tilde{b}_2] = \frac{1}{p_0 l^2}[a_1 \quad b_1 \quad a_2 \quad b_2], \tag{2.126}$$

we find that the conditions (2.125) assume the form

$$\tilde{\mathscr{W}}^{\mathrm{m}} + \sin\beta\tan\beta\left[\left(\operatorname{ber} x - \frac{x}{2}\operatorname{ber}' x - \frac{2}{x}\operatorname{bei}' x\right)\tilde{a}_2 + \left(\operatorname{bei} x - \frac{x}{2}\operatorname{bei}' x + \frac{2}{x}\operatorname{ber}' x\right)\tilde{b}_2\right] = 0,$$

$$\frac{2}{x^2}\left[\left(2\operatorname{bei} x + \frac{4}{x}\operatorname{ber}' x - x\operatorname{bei}' x\right)\tilde{a}_2 + \left(-2\operatorname{ber} x + \frac{4}{x}\operatorname{bei}' x + x\operatorname{ber}' x\right)\tilde{b}_2\right] = 0, \tag{2.127}$$

where $\upsilon = 1$ and $x = 32$ at the lower edge, and $\tilde{\mathscr{W}}^{\mathrm{m}}$ is given by (2.123). In order to solve these equations, we need the values of the Kelvin functions for $x = 32$. This value of the argument is outside the interval covered by tables, and the values of the functions are therefore computed by means of the asymptotic formulae 9.10.9 to 9.10.26 in Abramowitz and Stegun,[38] using an HP-67 programmable pocket calculator. If we insert the numerical values in (2.127) and divide by $\sin\beta\tan\beta$ in $(2.127)_1$ and by $2/x^2$ in $(2.127)_2$, we obtain the linear equations

$$-3.384\,04 \times 10^9\tilde{a}_2 - 6.341\,47 \times 10^9\tilde{b}_2 = 0.866\,025,$$
$$12.6829 \times 10^9\tilde{a}_2 - 6.768\,09 \times 10^9\tilde{b}_2 = 0,$$

the solution of which is given by

$$\tilde{a}_2 = -0.567\,235 \times 10^{-10}, \qquad \tilde{b}_2 = -1.062\,956 \times 10^{-10}. \tag{2.128}$$

In order to formulate the edge conditions at the *upper edge*, we shall use the results derived in section 2.2.4 concerning ring beams. Referring to Fig. 23, it will be seen that the conditions at the upper edge of the conical shell correspond to the situation in which the upper shell segment in Fig. 23 is missing. The fact that we have to do with a single shell segment in the present case implies that the two conditions $(2.44)_{1,2}$ that express the displacement continuity between the segments should be omitted, and the last condition $(2.44)_4$ should be replaced by the condition $M_{11} = 0$. Moreover, the condition $(2.44)_3$ is simplified because $\mathscr{H}_1 = \bar{P}_h = 0$ in the present case. We therefore obtain the two edge conditions

$$\mathscr{H} = \frac{E_b A}{r_0^2} \mathscr{W}, \qquad M_{11} = 0. \tag{2.129}$$

Fig. 29 shows the forces acting on the edge of the shell and on the ring beam, the indices m and h denoting quantities associated with the membrane solution and the solution of the homogeneous equations, respectively.

If we introduce the dimensionless quantities (2.122), the conditions (2.129) can be written in the form

$$\tilde{\mathscr{W}} = \frac{r_0^2 h}{Al} \tilde{\mathscr{H}}, \qquad \tilde{M}_{11} = 0, \tag{2.130}$$

in which the Young's moduli of the shell and the beam are assumed to have the same value. In this case, we shall neglect the terms with a_2 and b_2 in the representation (2.82) of the solution of the homogeneous equations (these terms correspond to the edge effect originating from the lower edge). The conditions (2.130) can then be written in the following manner in terms of the dimensionless quantities (2.122) and the dimensionless arbitrary constants (2.126):

$$\begin{aligned}
&\tilde{\mathscr{W}}^{\mathrm{m}} + \sin\beta\tan\beta\left[\left(\ker x - \frac{x}{2}\ker' x - \frac{2}{x}\operatorname{kei}' x\right)\tilde{a}_1\right.\\
&\qquad \left. + \left(\operatorname{kei} x - \frac{x}{2}\operatorname{kei}' x + \frac{2}{x}\ker' x\right)\tilde{b}_1\right]\\
&= \frac{r_0^2 h}{Al}\left\{\tilde{\mathscr{H}}^{\mathrm{m}} + \frac{1}{v\cos\beta}\left[\left(-\ker x + \frac{2}{x}\operatorname{kei}' x\right)\tilde{a}_1\right.\right.\\
&\qquad\qquad \left.\left. - \left(\operatorname{kei} x + \frac{2}{x}\ker' x\right)\tilde{b}_1\right]\right\},
\end{aligned} \tag{2.131}$$

$$\begin{aligned}
&\frac{2}{x^2}\left[\left(2\operatorname{kei} x + \frac{4}{x}\ker' x - x\operatorname{kei}' x\right)\tilde{a}_1\right.\\
&\qquad \left. + \left(-2\ker x + \frac{4}{x}\operatorname{kei}' x + x\ker' x\right)\tilde{b}_1\right] = 0,
\end{aligned}$$

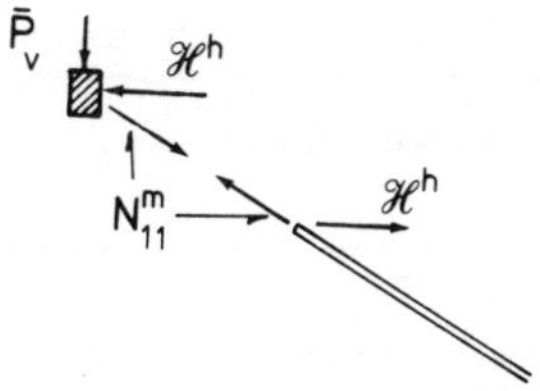

Fig. 29. Forces acting on edge of shell and on ring beam

where $\upsilon = 0.3$ and $x = 17.5271$ at the upper edge.

Let the cross-sectional area of the ring beam be given by

$$A = 200 \text{ cm}^2,$$

and note that r_0 is given by

$$r_0 = l \sin \beta = 8.660\,25 \text{ m}.$$

Substituting the numerical values in (2.131), we obtain the following linear equations

$$3.856\,08 \times 10^{-5}\,\tilde{a}_1 + 0.130\,758 \times 10^{-5}\,\tilde{b}_1 = 0.568\,141,$$
$$1.078\,88 \times 10^{-5}\,\tilde{a}_1 - 2.137\,69 \times 10^{-5}\,\tilde{b}_1 = 0,$$

where the second equation (2.131) has been multiplied by the factor $x^2/2$. The solution of these equations is given by

$$\tilde{a}_1 = 0.144\,857 \times 10^5, \qquad \tilde{b}_1 = 0.073\,109 \times 10^5. \qquad (2.132)$$

In these and the following calculations, the values of the Kelvin functions are again computed with the help of the asymptotic formulae.

Contact forces and couples. Having determined the arbitrary constants, we can now calculate the values of the contact forces and couples in the shell. We add the membrane solution to the solution of the homogeneous equations (2.82), introduce dimensionless quantities, and insert the values (2.128) and (2.132) of the arbitrary constants. By computing the resulting expressions we obtain the values given in Table 2.1 and plotted graphically in Fig. 30. It will be seen that the values of the contact forces and couples in the central part of the shell agree closely with the values furnished by the membrane theory, while edge effects are significant in the vicinity of the edges. This kind of behaviour is typical of thin shell structures.

We shall finally estimate the order of magnitude of the error caused by omitting the contribution from the edge effect from the opposite edge. The order of magnitude of the functions ber x, bei x, ber′ x, and bei′ x is 10^8 at the lower edge ($x = 32$), and 10^4 at the upper edge ($x = 17.5271$). These functions are therefore reduced by a factor of about 10^{-4} in the transition from the lower to the upper edge. Similarly it is found that the functions ker x, kei x, ker′ x, and kei′ x are reduced by a factor that does not exceed 10^{-4} in the

Table 2.1 Values of contact forces and couples

υ	Solution of exact equations for conical shell $\tilde{N}_{11}$	$\tilde{N}_{22}$	$\tilde{Q}_1 \times 10^2$	$\tilde{M}_{11} \times 10^3$	Geckeler's method $\tilde{N}_{11}$	$\tilde{N}_{22}$	$\tilde{Q}_1 \times 10^2$	$\tilde{M}_{11} \times 10^3$
0.3	−0.209	−1.572	−5.248	0	−0.209	−1.574	−5.235	0
0.35	−0.357	−0.728	0.389	−0.691	−0.361	−0.730	0.647	−0.773
0.4	−0.411	−0.557	0.658	−0.278	−0.416	−0.531	0.897	−0.281
0.5	−0.501	−0.729	0.029	0.022	−0.499	−0.736	−0.043	0.027
0.7	−0.701	−1.092	0.046	−0.057	−0.702	−1.099	0.094	−0.029
0.8	−0.811	−1.317	0.660	0.250	−0.810	−1.300	0.561	0.271
0.9	−0.913	−1.147	0.729	1.047	−0.910	−1.144	0.592	0.988
0.95	−0.938	−0.687	−0.667	1.071	−0.938	−0.706	−0.670	1.030
1.00	−0.934	0	−3.834	0	−0.934	0	−3.827	0

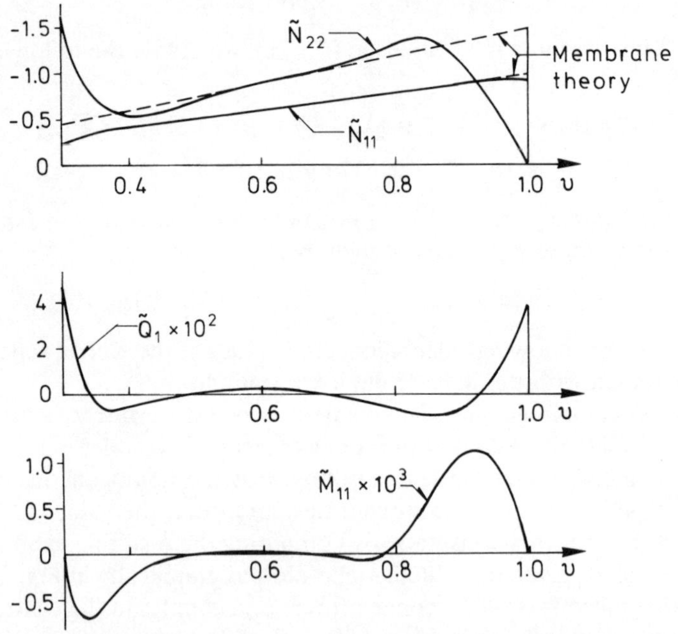

Fig. 30. Contact forces and couples in conical shell

transition from the upper to the lower edge. As the relative error of the solution of the homogeneous equations will be of the same order of magnitude as this damping factor, we conclude that the error is insignificant.

(b) *Geckeler's method*. We shall now determine an approximate solution for the conical shell by means of Geckeler's method (see section 2.6). Since

$$R_{2u} = l_1 \tan\beta, \qquad R_{2l} = l \tan\beta,$$

it follows from (2.113) that

$$\eta = \frac{1}{l_1 \tan\beta}(s - l_1) = \cot\beta\left(\frac{l}{l_1}\upsilon - 1\right),$$
$$\zeta = \frac{1}{l \tan\beta}(l - s) = \cot\beta(1 - \upsilon). \tag{2.133}$$

From (2.114a) and (2.116a) we get

$$\kappa_u^2 = (\sqrt{3})\frac{l_1}{h}\tan\beta, \qquad \kappa_l^2 = (\sqrt{3})\frac{l}{h}\tan\beta, \tag{2.134a}$$

or, on substituting the numerical values,

$$\kappa_u = 10.7331, \qquad \kappa_l = 19.5959. \tag{2.134b}$$

When formulating the edge conditions, we shall again neglect the influence of the edge effect from the opposite edge. We use the representation (2.118) and introduce the expressions (2.123) for the membrane forces. We also introduce the dimensionless quantities (2.122) and dimensionless arbitrary constants defined by

$$[\tilde{a}_1 \quad \tilde{b}_1 \quad \tilde{a}_2 \quad \tilde{b}_2] = \frac{1}{p_0 l}[a_1 \quad b_1 \quad a_2 \quad b_2]. \tag{2.135}$$

Since $f(0) = 1$, and $g(0) = 0$, we obtain the following results.

Lower edge. The conditions (2.125) assume the form

$$\sin^2\beta\tan\beta + \kappa_1\tan\beta\cos\beta\,(\tilde{a}_2 - \tilde{b}_2) = 0,$$
$$\tilde{a}_2 + \tilde{b}_2 = 0. \tag{2.136}$$

By inserting the numerical values, we obtain from these equations

$$\tilde{a}_2 = -\tilde{b}_2 = -0.038\,2733.$$

Upper edge. The conditions (2.130) assume the form

$$0.3^2\sin^2\beta\tan\beta + \kappa_u\tan\beta\cos\beta\,\frac{l_1}{l}(\tilde{b}_1 - \tilde{a}_1) = \frac{r_0^2 h}{Al}\left(\frac{1}{2}(\tan\beta)0.3 + \frac{\tilde{a}_1}{\cos\beta}\right)$$
$$\tilde{a}_1 + \tilde{b}_1 = 0. \tag{2.137}$$

By inserting the numerical values, we obtain from these equations

$$\tilde{a}_1 = -\tilde{b}_1 = -0.052\,3601.$$

By introducing these values of the arbitrary constants into (2.118), we can now calculate the values of the contact forces and couples in the shell. The results are given in Table 2.1. It will be seen that the agreement between the results obtained by solving the exact equations of the conical shell and the approximate results according to Geckeler's method is quite good. The maximum deviations between the two sets of results amount to 1.6% for $\tilde{N}_{11}$ and

$\tilde{N}_{22}$, 4.9% for $\tilde{Q}_1$, and 7.7% for $\tilde{M}_{11}$, the deviations being given as percentages of the relevant maximum values.

2.8 Concluding remarks

In the present chapter we have studied the analysis of shells of revolution under the assumption of complete rotational symmetry. It was found that the solution of the problem could be reduced to the integration of a system of two ordinary differential equations (see (2.25)).

If the load is no longer rotationally symmetrical, it is still possible to reduce the analytical solution to the integration of systems of ordinary differential equations. We shall briefly indicate how this is done. As the quantities A_1, A_2, R_1, and R_2 are independent of the angle θ (see Fig. 19 in section 2.2), the coefficients of the governing equations will likewise be independent of θ. If the load, the displacements, and the contact forces and couples are expanded as Fourier series in θ, it will be found that the variables φ and θ can be separated, and that, for each term of the Fourier series, a corresponding system of ordinary differential equations can be derived. A detailed description of this method, together with solutions for spherical and conical shells loaded by unsymmetric loads, can be found in Flügge's book.[4]

Problems

2.1 A hemispherical dome of constant thickness is clamped at the edge (i.e. for $\varphi = \pi/2$). It is assumed that Poisson's ratio is zero. The load is a constant vertical load per unit area of middle surface.

(1) Determine the contact forces N_{11} and N_{22}, the displacement $\mathscr{W}$, and the rotation χ according to the membrane theory.

(2) Use Geckeler's method to treat the edge effect. Determine the values of the arbitrary constants when $R/h = 100$. Find the values of M_{11} and N_{11} at the edge, and determine the maximum value of N_{22}.

2.2 The wall of a water tank consists of an elastic, circular cylindrical shell of constant thickness and with vertical axis. The shell is free at the upper edge ($z = 0$) and clamped at the lower edge ($z = L$).

Suppose that the tank is filled to its rim with water, and let

$$R/h = 12\sqrt{3}, \qquad L/R = 2, \qquad \nu = 0,$$

where R is the radius of the cylinder, h is the thickness, and ν is Poisson's ratio.

Determine the maximum bending moment M_{11} and the maximum ring force N_{22} due to the water pressure when the influence of the edge effect from the opposite edge is neglected.

2.3 An infinitely long, cylindrical tube (radius R) of constant thickness h is subjected to a uniform internal pressure p_0. The tube is provided with stiffening rings with spacing L, and both the shell and the stiffening rings are assumed to be elastic. Let

$$R/h = 49/\sqrt{3}, \qquad L/R = 0.5,$$
$$A = \tfrac{1}{2}hL, \qquad \nu = 0,$$

where A is the cross-sectional area of one ring, and ν is Poisson's ratio. It is assumed that the same value of Young's modulus applies to both the shell and the stiffening rings.

Determine the bending moment M_{11} and the ring force N_{22} at a stiffening ring and halfway between two adjacent stiffeners (use the symmetry of the problem to reduce the number of unknown arbitrary constants to two).

2.4 A cylindrical boiler drum is closed at the ends by spherical caps (see Fig. 2A). The boiler is subjected to a uniform internal pressure p_0.

The drum and the caps are regarded as elastic shells of constant thickness. Formulate the equations that express the conditions of transition at the junction between the cylindrical and the spherical shell when the influence of the edge effects from the opposite edges is neglected. Use Geckeler's method for the spherical shell.

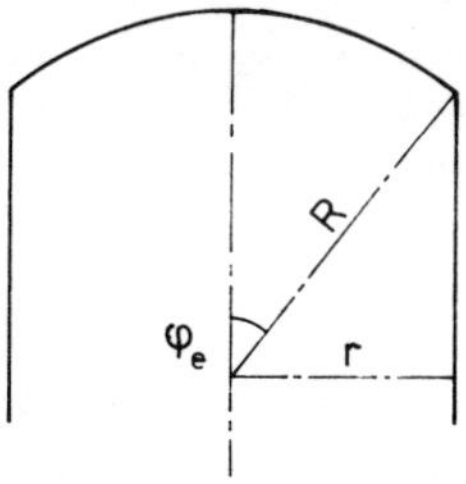

Fig. 2A

2.5 The curved walls of a pressure vessel are made up of a conical and a cylindrical shell, both of constant thickness (see Fig. 2B). The vessel is subjected to a uniform internal pressure p_0, and the shells are assumed to be elastic.

Formulate the equations that express the conditions of transition at the junction between the cylindrical and the conical shell when the edge effects from the opposite edges are neglected.

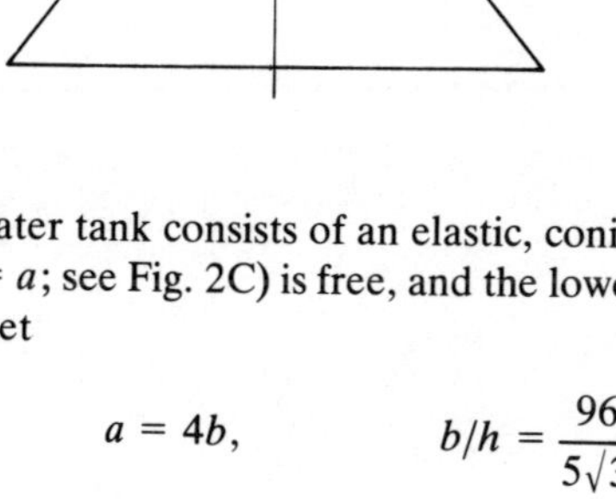

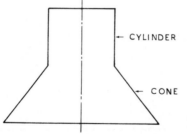

Fig. 2B

2.6 The wall of a water tank consists of an elastic, conical shell of constant thickness. The upper edge ($s = a$; see Fig. 2C) is free, and the lower edge ($s = b$, where $b < a$) is simply supported. Let

$$a = 4b, \qquad b/h = \frac{96}{5\sqrt{3}}$$

$$\tan \beta = \frac{15}{8}, \qquad \nu = 0,$$

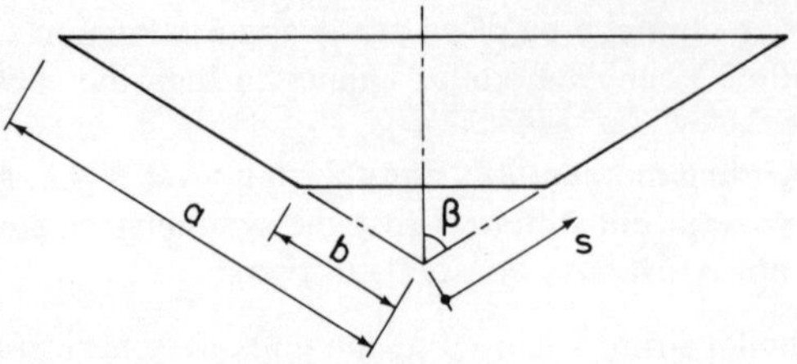

Fig. 2C

where h is the thickness, β is the angle between the axis of the cone and the generator, and ν is Poisson's ratio, and let the tank be filled to its rim.

(1) Determine the contact forces N_{11} and N_{22} and the displacement $\mathscr{W}$ according to the membrane theory.

(2) Use Geckeler's method to treat the edge effect at the lower edge, neglecting the influence of the edge effect from the opposite edge. Determine the arbitrary constants, and find N_{11} at the lower edge and the maximum values of N_{22} and M_{11}.

3

SHALLOW SHELLS

3.1 Simplifications in governing equations

Exact analytical solutions of the governing equations of the theory of shells can, in general, only be obtained for shells of simple geometrical shape. In the case of shells of more complicated shape, recourse must be had to numerical methods or to approximate analytical methods.

As far as *numerical methods* are concerned, powerful computer-oriented methods such as the finite element method and various types of finite difference methods are available for the calculation of shell structures. However, a detailed account of these numerical methods is beyond the scope of the present book, and the reader is referred to the literature on these topics (see, for example, Zienkiewicz[44] on the finite element method, and Kraus[9] on finite difference methods).

Approximate analytical solutions can be obtained by introducing *simplifications in the governing equations*. The resulting approximate theories are often useful from a practical point of view because of the relative ease with which solutions can be obtained. Moreover, several problems that cannot be solved analytically by means of the exact theory become amenable to analytical treatment when approximate theories are used. In the present chapter, we shall investigate two types of simplifications in the governing equations, and we shall direct our attention particularly to the theory of shallow shells.

Simplification 1. Let us consider the expressions (1.96) for the bending measures. Now, for several types of shell problems it can be shown that the contributions from the tangential displacements v_1 and v_2 in the formulae for the bending measures can be omitted, to a good approximation. This means that the following simplified expressions are used so that the bending measures are assumed to be determined solely by the normal component v_3 of the displacement.

$$\begin{aligned} \kappa_{11} &= -\frac{1}{A_1}\left(\frac{v_{3,1}}{A_1}\right)_{,1} - \frac{A_{1,2}}{A_1 A_2^2} v_{3,2}, \\ \kappa_{22} &= -\frac{1}{A_2}\left(\frac{v_{3,2}}{A_2}\right)_{,2} - \frac{A_{2,1}}{A_1^2 A_2} v_{3,1}, \end{aligned} \tag{3.1}$$

$$\kappa_{12} = \kappa_{21} = -\frac{1}{A_1A_2}\left(v_{3,12} - \frac{A_{1,2}}{A_1}v_{3,1} - \frac{A_{2,1}}{A_2}v_{3,2}\right), \tag{3.1}$$

The expressions $(1.83)_{1,2}$ for the components ω_1 and ω_2 of the rotation vector are simplified in a similar manner by omitting the contributions from v_1 and v_2. The application of the simplified expressions for the bending measures and the rotations implies, as will now be explained, that the equations of equilibrium should be simplified correspondingly. We shall stipulate that the principle of virtual work (1.127) and the constitutive equations (1.135) shall be valid in our approximate theory, where the bending measures $\kappa_{\alpha\beta}$ are given by the simplified expressions (3.1), the strain measures $\varepsilon_{\alpha\beta}$ are still given by (1.66), and the components of the rotation vector $\boldsymbol{\omega}$ are given by the simplified version of (1.83). If we now transform the internal work in the approximate principle of virtual work by means of Green's theorem, it can be shown that this principle is equivalent to a set of approximate equilibrium equations together with approximate expressions for the reduced edge loads. The approximate equations of equilibrium have the form:

$$\begin{aligned}
&(A_2N_{11})_{,1} + (A_1N_{12})_{,2} + A_{1,2}N_{12} - A_{2,1}N_{22} + A_1A_2p_1 = 0,\\
&(A_1N_{22})_{,2} + (A_2N_{12})_{,1} + A_{2,1}N_{12} - A_{1,2}N_{11} + A_1A_2p_2 = 0,\\
&\frac{N_{11}}{R_1} + \frac{N_{22}}{R_2} + \frac{1}{A_1A_2}[(A_2Q_1)_{,1} + (A_1Q_2)_{,2}] + p_3 = 0, \qquad (3.2)\\
&Q_\alpha = \frac{1}{A_1A_2}[(A_\beta M_{\alpha\alpha})_{,\alpha} + (A_\alpha M_{\alpha\beta})_{,\beta} + A_{\alpha,\beta}M_{\alpha\beta} - A_{\beta,\alpha}M_{\beta\beta}]. \qquad \alpha \neq \beta.
\end{aligned}$$

In these equations, $N_{\alpha\beta}$ and $M_{\alpha\beta}$ are the effective contact quantities, and Q_α is the effective transverse shear force, since the last equation (3.2) is identical with (1.122). It will now be seen that the approximate equations of equilibrium (3.2) differ in the following ways from the complete equations of equilibrium (1.111) written in terms of the effective contact forces and couples. The contributions from the transverse shear forces Q_α have vanished in the first two equations (3.2), and the contact forces and couples appearing in (3.2) are now the effective contact quantities $N_{\alpha\beta}$ and $M_{\alpha\beta}$, both of which are symmetrical with respect to the indices α and β. We also note that the use of the principle of virtual work as the basis of our derivation of the equilibrium conditions means that the resulting set of equilibrium equations does not include the equation of moment equilibrium about the normal $(1.111b)_3$ since this equation cannot be derived from the present version of the principle of virtual work. It would be possible to introduce equation $(1.111b)_3$ as an additional condition, in which case the skew-symmetric part of $N^{u}_{\alpha\beta}$ could be determined (this would lead to the relation (1.121)), but in the present approximate theory it is unnecessary to introduce the unsymmetric system $N^{u}_{\alpha\beta}$.

It should also be noted that it can be shown by means of the principle of virtual work and the constitutive equations (which are valid by assumption in the present case) that the theorem on the uniqueness of the solution, Betti's theorem, and the usual energy principles all retain their validity in the present approximate theory.

So far as the validity of simplification 1 is concerned, it can be shown, by means of methods similar to those employed by Koiter,[24] that the results obtained by this simplification will, in general, be acceptable if one of the following sets of conditions are satisfied:

(1) $$\frac{\kappa h}{2\varepsilon} = O(1) \wedge \left(\frac{L}{R}\right)^2 \ll 1,$$

(2) $$\frac{\kappa h}{2\varepsilon} \ll 1 \wedge 1 \ll \frac{L^2}{hR} \leq O\left(\frac{L}{h}\right).$$

In these expressions, ε and κ denote the numerically largest principal values of the strain and bending measures, respectively (so that $\kappa h/2\varepsilon$ is the ratio of the numerically largest flexural strain to the numerically largest extensional strain), L denotes the 'wavelength' of the deformation pattern (see (1.132a)), and R denotes the numerically smallest principal radius of curvature of the middle surface.

It will be seen that the flexural stresses are of the same order of magnitude as the membrane stresses in the first case, whereas the membrane stresses dominate in the second case.

We shall now turn to an approximate method that can be used for the analysis of *shallow shells*. A shell is said to be shallow when its middle surface does not deviate much from a plane. More precisely, a shallow shell can be defined in the following manner. Suppose that an orthogonal, Cartesian coordinate system X, Y, Z is given in space, and let the middle surface of the shell be given by an equation of the form (see Fig. 31)

$$z = z(x, y). \tag{3.3}$$

The shell is said to be shallow when the conditions

$$(\partial z/\partial x)^2 \ll 1, \qquad (\partial z/\partial y)^2 \ll 1 \tag{3.4}$$

are satisfied everywhere on the middle surface.

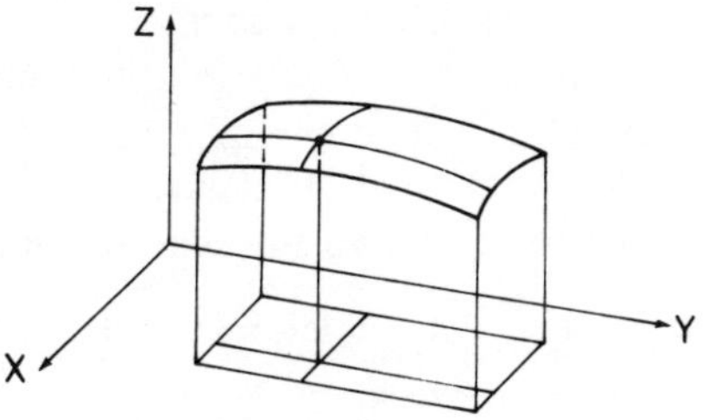

Fig. 31. Shallow shell

Simplification 2. Assuming that we are dealing with a shallow shell, the following simplification will be introduced. In the governing equations, we neglect squares and products of the derivatives $\partial z/\partial x$ and $\partial z/\partial y$ compared with unity.

When analysing shallow shells, it is convenient to let the middle surface be represented by an equation of the form (3.3). We shall therefore employ the following type of parametric representation for the middle surface

$$x = \theta_1, \qquad y = \theta_2, \qquad z = z(\theta_1, \theta_2), \tag{3.5}$$

so that the projections of the θ_1- and θ_2-curves on the XY-plane form straight lines parallel to the X- and Y-axes, respectively. It follows that the coordinate curves on the middle surface will, in general, be neither mutually orthogonal nor lines of curvature.

We shall now apply simplification 2 to derive approximate expressions for the coefficients of the first and second fundamental forms of the middle surface. By means of (1.4) and (1.11), we have

$$\begin{aligned} &\mathbf{a}_1 = [1 \quad 0 \quad z_{,1}], \qquad \mathbf{a}_2 = [0 \quad 1 \quad z_{,2}], \\ &\mathbf{a}_3 = \frac{\mathbf{a}_1 \times \mathbf{a}_2}{|\mathbf{a}_1 \times \mathbf{a}_2|} = \frac{[-z_{,1} \quad -z_{,2} \quad 1]}{(1 + z_{,1}^2 + z_{,2}^2)^{1/2}} \simeq [-z_{,1} \quad -z_{,2} \quad 1]. \end{aligned} \tag{3.6}$$

Further, from (1.9), (1.21) and (1.18), we have

$$\begin{aligned} &[a_{\alpha\beta}] = [\mathbf{a}_\alpha \cdot \mathbf{a}_\beta] = \begin{bmatrix} (1 + z_{,1}^2) & z_{,1}z_{,2} \\ z_{,1}z_{,2} & (1 + z_{,2}^2) \end{bmatrix} \simeq \begin{bmatrix} 1 & 0 \\ 0 & 1 \end{bmatrix}, \\ &A_1 = (a_{11})^{1/2} \simeq 1, \qquad A_2 = (a_{22})^{1/2} \simeq 1, \\ &b_{\alpha\beta} = \mathbf{a}_3 \cdot \mathbf{a}_{\alpha,\beta} = z_{,\alpha\beta}/(1 + z_{,1}^2 + z_{,2}^2)^{1/2} \simeq z_{,\alpha\beta}. \end{aligned} \tag{3.7}$$

It will now be seen that $a_{12} \simeq 0$, i.e. the coordinate curves can be regarded as mutually orthogonal in our approximate theory. However, it should be noted that these curves will not, in general, be lines of curvature in the approximate theory, since $b_{12} \simeq z_{12}$, and this quantity is generally not zero (cf. the conditions (1.25)).

In the derivation of the general theory of shells in Chapter 1, it was assumed that the coordinate curves were the lines of curvature. It follows that we cannot, in general, use the governing equations of the general theory in the curvilinear coordinate system of the shallow shell. The correct form of the governing equations in this case is most easily derived by means of tensor methods, but can alternatively be found in the following manner. Since $A_\alpha \simeq 1$, it follows from (1.27) that $\mathbf{e}_\alpha \simeq \mathbf{a}_\alpha$. Using this and the previous approximations for shallow shells, we can derive formulae for the derivatives of a vector corresponding to (1.40). In this way we find

$$\begin{aligned} &\mathbf{V}_{,1} = (V_{1,1} - z_{,11}V_3)\mathbf{e}_1 + (V_{2,1} - z_{,12}V_3)\mathbf{e}_2 + (V_{3,1} + z_{,11}V_1 + z_{,12}V_2)\mathbf{e}_3, \\ &\mathbf{V}_{,2} = (V_{1,2} - z_{,12}V_3)\mathbf{e}_1 + (V_{2,2} - z_{,22}V_3)\mathbf{e}_2 + (V_{3,2} + z_{,12}V_1 + z_{,22}V_2)\mathbf{e}_3. \end{aligned} \tag{3.8}$$

We now use the vector equations (1.64) and (1.95) for the strain and bending measures, and the equations of equilibrium (1.110) in vector form, since it can be shown that these equations are valid for arbitrary orthogonal curvilinear coordinates on the middle surface (and not just when the coordinate curves are the lines of curvature). The derivatives of the displacement vector **v** and of the contact force and couple vectors $\mathbf{N}_\alpha$, $\mathbf{M}_\alpha$ are calculated with the help of (3.8), and we also introduce the above simplification 1 (the contributions of the tangential displacements to the bending measures are neglected, and the equations of equilibrium are modified accordingly). As a result of these calculations, the details of which will be omitted, the following formulae are obtained:

$$\begin{aligned}
&\varepsilon_{11} = v_{1,1} - z_{,11}v_3, \qquad \varepsilon_{22} = v_{2,2} - z_{,22}v_3,\\
&\varepsilon_{12} = \varepsilon_{21} = \tfrac{1}{2}(v_{1,2} + v_{2,1}) - z_{,12}v_3, \\
&\kappa_{11} = -v_{3,11}, \qquad \kappa_{22} = -v_{3,22}, \qquad \kappa_{12} = \kappa_{21} = -v_{3,12}.
\end{aligned} \tag{3.9}$$

$$\begin{aligned}
&N_{11,1} + N_{12,2} + p_1 = 0, \qquad N_{12,1} + N_{22,2} + p_2 = 0,\\
&Q_{1,1} + Q_{2,2} + N_{11}z_{,11} + 2N_{12}z_{,12} + N_{22}z_{,22} + p_3 = 0,\\
&Q_\alpha = M_{\alpha\alpha,\alpha} + M_{\alpha\beta,\beta} \qquad (\alpha \neq \beta),
\end{aligned} \tag{3.10}$$

and we have $N_{\alpha\beta} = N_{\beta\alpha}$ and $M_{\alpha\beta} = M_{\beta\alpha}$. The principle of virtual work (1.127) is valid with $\varepsilon_{\alpha\beta}$ and $\kappa_{\alpha\beta}$ being given by (3.9). Using simplifications 1 and 2, we find for the components of the rotation vector

$$\omega_1 = -v_{3,1}, \qquad \omega_2 = -v_{3,2}, \qquad \omega_3 = \tfrac{1}{2}(v_{2,1} - v_{1,2}), \tag{3.11a}$$

and the following formulae, corresponding to (1.40), are obtained for the reduced edge loads, assuming that the boundary arc is part of a θ_2-curve

$$\begin{aligned}
&\mathrm{N}_{11} = N_{11}, \qquad \mathrm{N}_{12} = N_{12}, \qquad \mathrm{M}_{11} = M_{11},\\
&\mathrm{Q}_1 + \mathrm{M}_{12,2} = Q_1 + M_{12,2}.
\end{aligned} \tag{3.11b}$$

Equations (3.9) and (3.10), supplemented by (3.11) and the constitutive equations (1.135), constitute the *governing equations* of the theory of *shallow shells*.

We finally consider a special type of middle surface geometry for which the equations of shallow shell theory can be derived by means of the equations of the general theory of Chapter 1. Let the middle surface be a *translation surface* given by the equations

$$x = \theta_1, \qquad y = \theta_2, \qquad z = f(\theta_1) + g(\theta_2).$$

For such a surface we have the relation $b_{12} \simeq z_{,12} = 0$ (see (3.7)). From this and the previous result $a_{12} = 0$, we conclude that the coordinate curves of the translation surface can be regarded as lines of curvature in our approximate theory. We can therefore apply the equations of Chapter 1 when these are simplified according to simplifications 1 and 2. If we now insert the approximate values $A_1 = A_2 = 1$, $1/R_1 = z_{,11}$, and $1/R_2 = z_{,22}$ in (1.66), (3.1), and (3.2), we obtain precisely the equations of shallow shell theory with $z_{,12} = 0$.

3.2 Derivation of governing differential equations

In the following derivation of the differential equations of shallow shells, it will be assumed that the shell is of constant thickness. We shall also introduce the following changes of notation

$$v_3 = w, \qquad p_3 = p, \tag{3.12}$$

since the normal components of the displacement and the load will play a special role in the present method.

It will be convenient to use differential operators and matrix notation in the following derivation. If we put

$$\theta_1 = x, \qquad \theta_2 = y$$

and introduce the differential operators

$$\partial_x(\ ) = \frac{\partial}{\partial x}(\ ), \qquad \partial_y(\ ) = \frac{\partial}{\partial y}(\ ), \tag{3.13}$$

we find that the governing equations (3.9) and (3.10) can be written in matrix form as follows:

$$\begin{bmatrix} \varepsilon_{11} \\ \varepsilon_{22} \\ \varepsilon_{12} \end{bmatrix} = \begin{bmatrix} \partial_x & 0 & -z_{,xx} \\ 0 & \partial_y & -z_{,yy} \\ \frac{1}{2}\partial_y & \frac{1}{2}\partial_x & -z_{,xy} \end{bmatrix} \begin{bmatrix} v_1 \\ v_2 \\ w \end{bmatrix},$$

$$\begin{bmatrix} \kappa_{11} \\ \kappa_{22} \\ \kappa_{12} \end{bmatrix} = -\begin{bmatrix} \partial_x^2 \\ \partial_y^2 \\ \partial_x\partial_y \end{bmatrix} w, \tag{3.14}$$

$$\begin{bmatrix} \partial_x & 0 & \partial_y \\ 0 & \partial_y & \partial_x \end{bmatrix} \begin{bmatrix} N_{11} \\ N_{22} \\ N_{12} \end{bmatrix} + \begin{bmatrix} p_1 \\ p_2 \end{bmatrix} = \begin{bmatrix} 0 \\ 0 \end{bmatrix}, \tag{3.15}$$

$$[\partial_x \quad \partial_y] \begin{bmatrix} Q_1 \\ Q_2 \end{bmatrix} + [z_{,xx} \quad z_{,yy} \quad 2z_{,xy}] \begin{bmatrix} N_{11} \\ N_{22} \\ N_{12} \end{bmatrix} + p = 0, \tag{3.16}$$

$$\begin{bmatrix} \partial_x & 0 & \partial_y \\ 0 & \partial_y & \partial_x \end{bmatrix} \begin{bmatrix} M_{11} \\ M_{22} \\ M_{12} \end{bmatrix} = \begin{bmatrix} Q_1 \\ Q_2 \end{bmatrix}. \tag{3.17}$$

The constitutive equations can be written in the form

$$\begin{bmatrix} M_{11} \\ M_{22} \\ M_{12} \end{bmatrix} = D \begin{bmatrix} 1 & \nu & 0 \\ \nu & 1 & 0 \\ 0 & 0 & (1-\nu) \end{bmatrix} \begin{bmatrix} \kappa_{11} \\ \kappa_{22} \\ \kappa_{12} \end{bmatrix},$$

$$\begin{bmatrix} \varepsilon_{11} \\ \varepsilon_{22} \\ \varepsilon_{12} \end{bmatrix} = \frac{1}{Eh} \begin{bmatrix} 1 & -\nu & 0 \\ -\nu & 1 & 0 \\ 0 & 0 & (1+\nu) \end{bmatrix} \begin{bmatrix} N_{11} \\ N_{22} \\ N_{12} \end{bmatrix}, \tag{3.18}$$

where

$$D = \frac{Eh^3}{12(1-\nu^2)} \tag{3.18a}$$

is the bending stiffness. Equation $(3.18)_1$ represents the last three equations (1.135), while equation $(3.18)_2$ is obtained by solving the first three equations (1.135) with respect to $\varepsilon_{\alpha\beta}$.

We now represent $N_{\alpha\beta}$ in the following manner in terms of a *stress function* ψ:

$$\begin{bmatrix} N_{11} \\ N_{22} \\ N_{12} \end{bmatrix} = \begin{bmatrix} \partial_y^2 \\ \partial_x^2 \\ -\partial_x\partial_y \end{bmatrix} \psi - \begin{bmatrix} B_x \\ B_y \\ 0 \end{bmatrix}, \tag{3.19}$$

where

$$B_x = \int p_1 \, \mathrm{d}x, \qquad B_y = \int p_2 \, \mathrm{d}y. \tag{3.19a}$$

Substituting this expression for $N_{\alpha\beta}$ in (3.15), it will be seen that these two equations of equilibrium (projection in the tangential directions) are satisfied identically.

We proceed to derive two coupled differential equations for ψ and w. We premultiply both sides of $(3.14)_1$ by the vector $[\partial_y^2 \;\; \partial_x^2 \;\; -2\partial_x\partial_y]$ and obtain

$$[\partial_y^2 \;\; \partial_x^2 \;\; -2\partial_x\partial_y] \begin{bmatrix} \varepsilon_{11} \\ \varepsilon_{22} \\ \varepsilon_{12} \end{bmatrix} = -(\partial_y^2 z_{,xx} + \partial_x^2 z_{,yy} - 2\partial_x\partial_y z_{,xy})w, \tag{3.20}$$

since the terms involving v_1 and v_2 cancel out. The right-hand side of (3.20) should be interpreted in the following manner:

$$-(\partial_y^2(z_{,xx}w) + \partial_x^2(z_{,yy}w) - 2\partial_x\partial_y(z_{,xy}w))$$

i.e. the differential operators act on the following factors. Performing the differentiations, we find that several terms cancel out, and that only the expression

$$-(z_{,xx}w_{,yy} + z_{,yy}w_{,xx} - 2z_{,xy}w_{,xy})$$

remains. Equation (3.20) can therefore be written in the form

$$[\partial_y^2 \quad \partial_x^2 \quad -2\partial_x\partial_y]\begin{bmatrix}\varepsilon_{11}\\ \varepsilon_{22}\\ \varepsilon_{12}\end{bmatrix} = -(z_{,xx}\partial_y^2 + z_{,yy}\partial_x^2 - 2z_{,xy}\partial_x\partial_y)w. \quad (3.21)$$

It follows from $(3.14)_2$ that the right-hand side of (3.21) can be written $(z_{,xx}\kappa_{yy} + z_{,yy}\kappa_{xx} - 2z_{,xy}\kappa_{xy})$, and it can now be seen that (3.21) corresponds to the third equation of compatibility (1.104), since $A_1 = A_2 = 1$, $1/R_1 = z_{,xx}$, $1/R_2 = z_{,yy}$ for shallow shells, and $z_{,xy} = 0$ when the coordinate curves are the lines of curvature.

We shall now introduce the following *differential operators*:

$$\begin{aligned}\nabla^2 F &= \frac{\partial^2 F}{\partial x^2} + \frac{\partial^2 F}{\partial y^2},\\ \nabla^4 F &= \frac{\partial^4 F}{\partial x^4} + 2\frac{\partial^4 F}{\partial x^2 \partial y^2} + \frac{\partial^4 F}{\partial y^4},\\ L[F] &= \frac{\partial^2 z}{\partial x^2}\frac{\partial^2 F}{\partial y^2} - 2\frac{\partial^2 z}{\partial x \partial y}\frac{\partial^2 F}{\partial x \partial y} + \frac{\partial^2 z}{\partial y^2}\frac{\partial^2 F}{\partial x^2}.\end{aligned} \quad (3.22)$$

Substituting the expression (3.18) for $\varepsilon_{\alpha\beta}$ in (3.21), we obtain

$$[\partial_y^2 \quad \partial_x^2 \quad -2\partial_x\partial_y]\frac{1}{Eh}\begin{bmatrix}1 & -\nu & 0\\ -\nu & 1 & 0\\ 0 & 0 & (1+\nu)\end{bmatrix}\begin{bmatrix}N_{11}\\ N_{22}\\ N_{12}\end{bmatrix} + L[w] = 0,$$

and inserting the formula (3.19) for $N_{\alpha\beta}$ we further derive

$$[\partial_y^2 \quad \partial_x^2 \quad -2\partial_x\partial_y]\frac{1}{Eh}\left(\begin{bmatrix}\partial_y^2 - \nu\partial_x^2\\ \partial_x^2 - \nu\partial_y^2\\ -(1+\nu)\partial_x\partial_y\end{bmatrix}\psi - \begin{bmatrix}B_x - \nu B_y\\ B_y - \nu B_x\\ 0\end{bmatrix}\right) + L[w] = 0,$$

or

$$(\partial_x^4 + \partial_y^4 + 2\partial_x^2\partial_y^2)\psi - [B_{x,yy} + B_{y,xx} - \nu(B_{y,yy} + B_{x,xx})] + EhL[w] = 0. \quad (3.23)$$

Using (3.22), equation (3.23) can be written in the form

$$\nabla^4\psi + EhL[w] = (1+\nu)(B_{x,yy} + B_{y,xx}) - \nu\nabla^2(B_x + B_y). \quad (3.24)$$

This is one differential equation for ψ and w.

In order to find the second equation, we first eliminate the transverse shear forces from (3.16) and (3.17). We premultiply both sides of (3.17) by the vector $[\partial_x \quad \partial_y]$ and then substitute in (3.16). Thus

$$[\partial_x^2 \quad \partial_y^2 \quad 2\partial_x\partial_y]\begin{bmatrix}M_{11}\\ M_{22}\\ M_{12}\end{bmatrix} + [z_{,xx} \quad z_{,yy} \quad 2z_{,xy}]\begin{bmatrix}N_{11}\\ N_{22}\\ N_{12}\end{bmatrix} + p = 0.$$

Inserting the expressions (3.19) for $N_{\alpha\beta}$ and $(3.18)_1$ for $M_{\alpha\beta}$ (with $\kappa_{\alpha\beta}$ given by $(3.14)_2$), we get

$$-[\partial_x^2 \quad \partial_y^2 \quad 2\partial_x\partial_y]D\begin{bmatrix} \partial_x^2+\nu\partial_y^2 \\ \partial_y^2+\nu\partial_x^2 \\ (1-\nu)\partial_x\partial_y \end{bmatrix} w + [z_{,xx} \quad z_{,yy} \quad 2z_{,xy}]\left(\begin{bmatrix} \partial_y^2 \\ \partial_x^2 \\ -\partial_x\partial_y \end{bmatrix}\psi - \begin{bmatrix} B_x \\ B_y \\ 0 \end{bmatrix}\right)$$

$+ p = 0,$

or

$$D\nabla^4 w - L[\psi] = p - z_{,xx}B_x - z_{,yy}B_y. \tag{3.25}$$

This is the second equation for ψ and w. Equations (3.24) and (3.25) constitute a system of two coupled linear partial differential equations for the determination of ψ and w.

Consider now the case in which z (see (3.3)) is a polynomial of the second degree in x and y, i.e.

$$z = a + c_1x + c_2y + \tfrac{1}{2}(k_{11}x^2 + 2k_{12}xy + k_{22}y^2), \tag{3.26}$$

where a, c_α, and $k_{\alpha\beta}$ are constants. We then have

$$z_{,xx} = k_{11}, \qquad z_{,yy} = k_{22}, \qquad z_{,xy} = k_{12},$$

$$L[F] = k_{11}\frac{\partial^2 F}{\partial y^2} - 2k_{12}\frac{\partial^2 F}{\partial x \partial y} + k_{22}\frac{\partial^2 F}{\partial x^2}, \tag{3.27}$$

and in this case (3.24) and (3.25) are equations with constant coefficients.

3.3 Shallow shell over a rectangular plan

Let the middle surface of the shallow shell be given by the equation

$$z = \frac{1}{2}\left(\frac{x^2}{R_x} + \frac{y^2}{R_y}\right), \tag{3.28}$$

where R_x and R_y are constants, and let the projection of the boundary curve on the XY-plane be a rectangle with sides parallel to the coordinate axes (see Fig. 32). In order to simplify the following derivation, it will be assumed that the following conditions are satisfied:

$$|R_x| \geq R_y > 0. \tag{3.29}$$

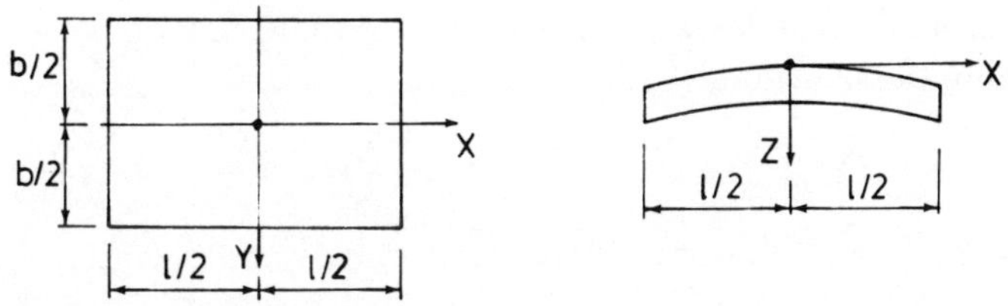

Fig. 32. Shallow shell over a rectangular plan

Depending on the sign of R_x, the equation (3.28) represents an elliptic paraboloid ($R_x > 0$), a parabolic cylinder ($1/R_x = 0$), or a hyperbolic paraboloid ($R_x < 0$).

We also assume that

$$p_1 = p_2 = 0, \qquad \text{i.e.} \qquad B_x = B_y = 0,$$

so that only the normal component p of the load differs from zero, and we assume that Poisson's ratio is zero ($\nu = 0$).

With these assumptions, the equations (3.24) and (3.25) assume the form

$$\nabla^4\psi + EhL[w] = 0, \tag{3.30}$$

$$D\nabla^4 w - L[\psi] = p, \tag{3.31}$$

where

$$L[\psi] = \frac{1}{R_x}\frac{\partial^2\psi}{\partial y^2} + \frac{1}{R_y}\frac{\partial^2\psi}{\partial x^2}, \tag{3.32}$$

and

$$D = \tfrac{1}{12}Eh^3 \tag{3.33}$$

since $\nu = 0$. Differentiating (3.30) and (3.31) we get

$$\nabla^8\psi + Eh\nabla^4 L[w] = 0, \qquad \frac{Eh^3}{12}L[\nabla^4 w] - L^2[\psi] = L[p],$$

where $L^2[\psi] = L[L[\psi]]$. Since we are dealing with differential operators with constant coefficients, these operators are permutable, so that $\nabla^4 L[w] = L[\nabla^4 w]$. We may therefore eliminate w so as to obtain the equation

$$\nabla^8\psi + \frac{12}{h^2}L^2[\psi] = -\frac{12}{h^2}L[p]. \tag{3.34}$$

This is an eighth order linear, partial differential equation with constant coefficients for ψ. In a similar manner, we may eliminate ψ to obtain the following equation for w:

$$\nabla^8 w + \frac{12}{h^2}L^2[w] = \frac{12}{Eh^3}\nabla^4 p. \tag{3.35}$$

We now express the contact forces and couples and the displacements in terms of ψ and w. We shall determine either the required functions themselves or partial derivatives with respect to x of these functions.

Introducing the notation

$$N_x = N_{11}, \qquad N_y = N_{22}, \qquad N_{xy} = N_{12},$$

$$Q_x = Q_1, \qquad Q_y = Q_2,$$

$$M_x = M_{11}, \qquad M_y = M_{22}, \qquad M_{xy} = M_{12},$$

we find with the help of equations (3.11) and (3.14) to (3.19) that

$$
\begin{aligned}
&N_x = \psi_{,yy}, \qquad N_{xy} = -\psi_{,xy}, \qquad N_y = \psi_{,xx}, \\
&M_x = -Dw_{,xx}, \qquad M_{xy} = -Dw_{,xy}, \qquad M_y = -Dw_{,yy}, \\
&Q_x = M_{x,x} + M_{xy,y} = -D(w_{,xxx} + w_{,xyy}), \\
&Q_y = M_{xy,x} + M_{y,y} = -D(w_{,xxy} + w_{,yyy}), \\
&S_x = Q_x + M_{xy,y} = -D(w_{,xxx} + 2w_{,xyy}), \qquad (3.36)\\
&S_y = Q_y + M_{xy,x} = -D(w_{,yyy} + 2w_{,xxy}), \\
&\frac{\partial v_1}{\partial x} = \frac{N_x}{Eh} + \frac{w}{R_x} = \frac{1}{Eh}\psi_{,yy} + \frac{w}{R_x}, \\
&\frac{\partial^2 v_2}{\partial x^2} = \frac{2}{Eh}\frac{\partial N_{xy}}{\partial x} - \frac{\partial^2 v_1}{\partial x \partial y} = \frac{-1}{R_x} w_{,y} - \frac{1}{Eh}(2\psi_{,xxy} + \psi_{,yyy}).
\end{aligned}
$$

$$\omega_2 = -w_{,y}.$$

The expressions for $\partial v_1/\partial x$ and $\partial^2 v_2/\partial x^2$ are obtained by means of the formulae

$$N_x = Eh\left(\frac{\partial v_1}{\partial x} - \frac{w}{R_x}\right) \quad \text{and} \quad N_{xy} = \tfrac{1}{2}Eh\left(\frac{\partial v_1}{\partial y} + \frac{\partial v_2}{\partial x}\right).$$

S_x and S_y denote reduced edge loads (see (3.11b)).

We now assume that the shell is subjected to the following *loading*:

$$p(x, y) = p_n(y) \cos \alpha_n x, \qquad (3.37a)$$

where

$$\alpha_n = n\pi/l, \qquad (3.37b)$$

and n is an odd integer ($n = 1, 3, 5, \ldots$). We seek a corresponding solution in the form

$$\psi(x, y) = \psi_n(y) \cos \alpha_n x, \qquad w(x, y) = w_n(y) \cos \alpha_n x. \qquad (3.38)$$

It will be shown in the following that the *edge conditions* for this solution at the edges $x = \pm l/2$ are given by

$$N_x = M_x = w = v_2 = 0, \qquad (3.39)$$

i.e. the shell is simply supported along these edges.

The solution (3.38) is determined as the sum of a particular integral and a solution of the homogeneous equations (each of these contributions being of the form (3.38)). As shown below, the trigonometric factors in (3.38) ensure that the edge conditions (3.39) are automatically satisfied at $x = \pm l/2$, and the solution of the homogeneous equations will now be determined in such a manner that the prescribed edge conditions at the edges $y = \pm$ b/2 are also satisfied.

A knowledge of the solution (3.38) will enable us to construct the solution for any load $p(x, y)$ that is symmetrical with respect to the Y-axis (i.e. for which $p(x, y) = p(-x, y)$). If this load is expanded in the X-direction in a Fourier half-range series of the form

$$p(x, y) = \sum_n p_n(y) \cos \alpha_n x, \qquad n = 1, 3, 5, \ldots \tag{3.40}$$

it will be seen that the corresponding solution is given by the series

$$\psi(x, y) = \sum_n \psi_n(y) \cos \alpha_n x, \qquad w(x, y) = \sum_n w_n(y) \cos \alpha_n x, \tag{3.41}$$

where $\psi_n(y)$ and $w_n(y)$ are the functions appearing in (3.38).

If we substitute the expressions (3.38) for ψ and w in the right-hand sides of (3.36), it will be seen that each of these statical and geometrical quantities will be given by an expression of the form

$$f(y) \cos \alpha_n x, \qquad \text{or} \qquad g(y) \sin \alpha_n x, \tag{3.42}$$

i.e. a function of y multiplied by either $\cos \alpha_n x$ or $\sin \alpha_n x$. The displacements v_1 and v_2 associated with this solution can also be written in the form (3.42), as will be proved in section 3.3.2.

In the following derivations, it will be convenient to introduce the dimensionless coordinate

$$\eta = \alpha_n y. \tag{3.43}$$

We note that the derivatives with respect to y and η are connected by the relation

$$\frac{\partial}{\partial y}(\ \ ) = \alpha_n \frac{\partial}{\partial \eta}(\ \ ). \tag{3.44}$$

It follows from the previous remarks that each of the required statical and geometrical quantities belonging to the solution (3.38) can be written in the form

$$F(x, \eta) = F_0 \tilde{F}(\eta) \varphi_{\text{trig}}(x), \tag{3.45}$$

where $\tilde{F}(\eta)$ is a *dimensionless function* of η, φ_{trig} is either $\cos \alpha_n x$ or $\sin \alpha_n x$, and F_0 is a constant with the same dimensions as F. The values of F_0 and φ_{trig} for the relevant statical and geometrical quantities are summarized in Table 3.1, where p_0 is a constant load intensity (force per unit area), and the appropriate trigonometric functions can be found by using equations (3.36) together with the fact that ψ and w vary in the x-direction as cosine functions.

It follows from (3.45) that the quantity F is completely determined once the corresponding dimensionless function $\tilde{F}(\eta)$ is known, and our objective in the following will therefore be to determine the dimensionless functions.

If we compare the expressions for p, ψ, and w from Table 3.1 and (3.45)

Table 3.1 F_0 and φ_{trig} values

F	F_0	φ_{trig}	F	F_0	φ_{trig}
p	p_0	$\cos \alpha_n x$	Q_x	$p_0 R_y \dfrac{h\alpha_n}{2\sqrt{3}}$	$\sin \alpha_n x$
ψ	$p_0 R_y/\alpha_n^2$	$\cos \alpha_n x$	Q_y	$p_0 R_y \dfrac{h\alpha_n}{2\sqrt{3}}$	$\cos \alpha_n x$
N_x	$p_0 R_y$	$\cos \alpha_n x$	S_y	$p_0 R_y \dfrac{h\alpha_n}{2\sqrt{3}}$	$\cos \alpha_n x$
N_y	$p_0 R_y$	$\cos \alpha_n x$	w	$\dfrac{p_0 R_y 2\sqrt{3}}{Eh^2\alpha_n^2}$	$\cos \alpha_n x$
N_{xy}	$p_0 R_y$	$\sin \alpha_n x$	v_1	$\dfrac{p_0 R_y}{Eh\alpha_n}$	$\sin \alpha_n x$
M_x	$p_0 R_y \dfrac{h}{2\sqrt{3}}$	$\cos \alpha_n x$	v_2	$\dfrac{p_0 R_y}{Eh\alpha_n}$	$\cos \alpha_n x$
M_y	$p_0 R_y \dfrac{h}{2\sqrt{3}}$	$\cos \alpha_n x$	ω_2	$\dfrac{p_0 R_y 2\sqrt{3}}{Eh^2\alpha_n}$	$\cos \alpha_n x$
M_{xy}	$p_0 R_y \dfrac{h}{2\sqrt{3}}$	$\sin \alpha_n x$			

with equations (3.37a) and (3.38), we obtain the relations

$$p_n(y) = p_0 \tilde{p}(\eta),$$

$$\psi_n(y) = \frac{p_0 R_y}{\alpha_n^2} \tilde{\psi}(\eta), \qquad w_n(y) = \frac{p_0 R_y 2\sqrt{3}}{Eh^2\alpha_{2n}^2} \tilde{w}(\eta), \tag{3.46}$$

where y and η are connected by (3.43).

We note that the factor $\cos \alpha_n x$ in the expressions for N_x, M_x, w, and v_2 ensures that the edge conditions (3.39) are satisfied at the edges $x = \pm l/2$.

We shall now introduce the dimensionless functions in (3.36). Let us consider equation $(3.36)_{11}$ for $\partial v_1/\partial x$ as an example. Using Table 3.1, we may write this equation in the form

$$\frac{p_0 R_y}{Eh} \tilde{v}_1(\eta) \cos \alpha_n x = \frac{p_0 R_y}{Eh} \tilde{\psi}''(\eta) \cos \alpha_n x + \frac{p_0 R_y}{Eh^2} \frac{2\sqrt{3}}{\alpha_n^2} \frac{\tilde{w}(\eta)}{R_x} \cos \alpha_n x,$$

or

$$\tilde{v}_1(\eta) = \tilde{\psi}''(\eta) + 2\gamma \tilde{w}(\eta),$$

where

$$\gamma = \frac{\sqrt{3}}{h\alpha_n^2 R_x}, \tag{3.47}$$

and primes denote differentiation with respect to η,

$$(\ )' = \frac{\mathrm{d}}{\mathrm{d}\eta}(\ ). \tag{3.48}$$

Treating the remaining equations (3.36) in a similar manner, we obtain the following relations between the dimensionless functions:

$$\begin{aligned}
\tilde{N}_x &= \tilde{\psi}'', \qquad \tilde{N}_{xy} = \tilde{\psi}', \qquad \tilde{N}_y = -\tilde{\psi},\\
\tilde{M}_x &= \tilde{w}, \qquad \tilde{M}_{xy} = \tilde{w}', \qquad \tilde{M}_y = -\tilde{w}'',\\
\tilde{Q}_x &= \tilde{w}'' - \tilde{w}, \qquad \tilde{Q}_y = \tilde{w}' - \tilde{w}''',\\
\tilde{S}_y &= -\tilde{w}''' + 2\tilde{w}',\\
\tilde{v}_1 &= \tilde{\psi}'' + 2\gamma\tilde{w}, \qquad \tilde{v}_2 = \tilde{\psi}''' - 2\tilde{\psi}' + 2\gamma\tilde{w}',\\
\tilde{\omega}_2 &= -\tilde{w}'.
\end{aligned} \tag{3.49}$$

3.3.1 The homogeneous equation and the auxiliary equation

The homogeneous equation (3.34) has the form

$$\left(\frac{\partial^2}{\partial x^2} + \frac{\partial^2}{\partial y^2}\right)^4 \psi + \frac{12}{h^2}\left(\frac{1}{R_x}\frac{\partial^2}{\partial y^2} + \frac{1}{R_y}\frac{\partial^2}{\partial x^2}\right)^2 \psi = 0. \tag{3.50}$$

Inserting the expression $(3.38)_1$ for ψ, we get

$$\left(\frac{\mathrm{d}^2}{\mathrm{d}y^2} - \alpha_n^2\right)^4 \psi_n^h + \frac{12}{h^2}\left(\frac{1}{R_x}\frac{\mathrm{d}^2}{\mathrm{d}y^2} - \frac{\alpha_n^2}{R_y}\right)^2 \psi_n^h = 0, \tag{3.51}$$

where the index h denotes the solution of the homogeneous equation. We now introduce the dimensionless function $\tilde{\psi}$ by means of $(3.46)_2$ and replace the independent variable y by η (see (3.43)). The resulting equation can be reduced to

$$\left(\frac{\mathrm{d}^2}{\mathrm{d}\eta^2} - 1\right)^4 \tilde{\psi}^{\mathrm{h}} + 4\gamma^2\left(\frac{\mathrm{d}^2}{\mathrm{d}\eta^2} - k\right)^2 \tilde{\psi}^{\mathrm{h}} = 0, \tag{3.52}$$

where the dimensionless constants γ and k are given by

$$\gamma = \frac{\sqrt{3}}{h\alpha_n^2 R_x}, \qquad k = \frac{R_x}{R_y}. \tag{3.52a}$$

Equation (3.52) is an ordinary, linear differential equation of the eighth order with constant coefficients for the determination of $\tilde{\psi}^{\mathrm{h}}$. As usual, we seek solutions of the form

$$\tilde{\psi}^{\mathrm{h}}(\eta) = C\,\mathrm{e}^{r\eta}. \tag{3.53}$$

The auxiliary equation. Substitution of (3.53) into (3.52) and cancellation of the common factor $C\,\mathrm{e}^{r\eta}$ yield the auxiliary equation

$$(r^2 - 1)^4 + 4\gamma^2(r^2 - k)^2 = 0. \tag{3.54}$$

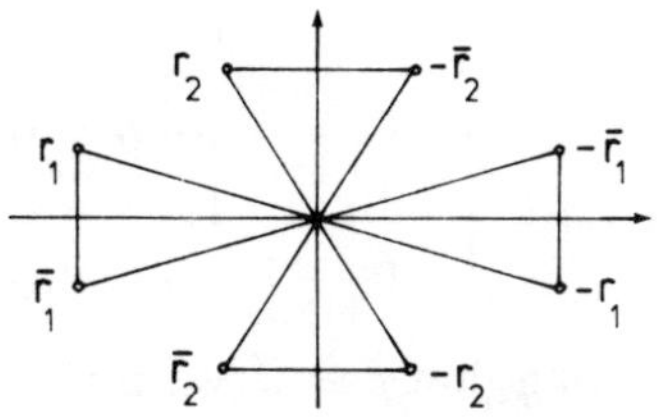

Fig. 33. Roots of auxiliary equation

This is an algebraic equation of the eighth degree. We observe that the coefficients are real and that the quantity r occurs in even powers only. This means that the roots occur in conjugate complex pairs, and also in pairs of the form $\pm r$. The eight roots can therefore be represented in the following manner (see also Fig. 33):

$$\pm r_1, \qquad \pm \bar{r}_1, \qquad \pm r_2, \qquad \pm \bar{r}_2,$$

so that equation (3.54) has only two essentially different roots r_1 and r_2.

Let us write (3.54) in the form

$$(r^2 - 1)^4 = -4\gamma^2(r^2 - k)^2.$$

Extracting the square root on both sides, we get

$$(r^2 - 1)^2 = \pm 2\mathrm{i}\gamma(r^2 - k), \tag{3.55}$$

or

$$r^4 - 2(1 \pm \mathrm{i}\gamma)r^2 + (1 \pm 2\mathrm{i}\gamma k) = 0,$$
$$r^2 = 1 \pm \mathrm{i}\gamma \oplus [-\gamma^2 \mp 2\mathrm{i}\gamma(k - 1)]^{1/2} = 0,$$

where the two sets of double signs not surrounded by circles should be read together, in the sense that either the two upper signs or the two lower signs are used. Let

$$\beta = 2\gamma(k - 1) = \frac{2\sqrt{3}}{h\alpha_n^2}\left(\frac{1}{R_y} - \frac{1}{R_x}\right) \geqslant 0, \tag{3.56}$$

where the last inequality follows from (3.29). We then obtain

$$r^2 = 1 \pm \mathrm{i}\gamma \oplus (-\gamma^2 \mp \mathrm{i}\beta)^{1/2}. \tag{3.57}$$

Now, the following formula is valid for the square root of an arbitrary complex number:

$$(a + \mathrm{i}b)^{1/2} = \pm\left(\frac{1}{\sqrt{2}}[a + (a^2 + b^2)^{1/2}]^{1/2} + \frac{\mathrm{i}}{\sqrt{2}}[-a + (a^2 + b^2)^{1/2}]^{1/2}(\operatorname{sgn} b)\right) \tag{3.58}$$

where

$$\operatorname{sgn} b = \begin{cases} +1 & \text{for } b > 0, \\ 0 & \text{for } b = 0, \\ -1 & \text{for } b < 0. \end{cases}$$

Using this formula and noting that $\beta \geqslant 0$, we find

$$\oplus(-\gamma^2 \mp i\beta)^{1/2} = \oplus(s \mp it), \tag{3.59}$$

where

$$s = \frac{1}{\sqrt{2}}[-\gamma^2 + (\gamma^4 + \beta^2)^{1/2}]^{1/2} \geqslant 0, \tag{3.60}$$

$$t = \frac{1}{\sqrt{2}}[\gamma^2 + (\gamma^4 + \beta^2)^{1/2}]^{1/2} \geqslant |\gamma|. \tag{3.61}$$

Inserting (3.59) in (3.57), we obtain

$$r^2 = 1 \oplus s \mp i(\oplus t - \gamma), \tag{3.62}$$

where the two sets of double signs surrounded by circles should be read together (either the two upper signs or the two lower signs).

We now denote the values of r^2 obtained by using the upper signs or the lower signs throughout equation (3.62) by r_1 and r_2, respectively. Hence

$$\begin{aligned} r_1^2 &= 1 + s - i(t - \gamma), \\ r_2^2 &= 1 - s - i(t + \gamma). \end{aligned} \tag{3.63}$$

Using (3.58) with the minus sign at the beginning of the right-hand side, and noting that $(t - \gamma) \geqslant 0$ and $(t + \gamma) \geqslant 0$, we obtain the following expressions for r_1 and r_2:

$$r_1 = -j_1 + ik_1, \qquad r_2 = -j_2 + ik_2, \tag{3.64}$$

where

$$\begin{aligned} j_1 &= \frac{1}{\sqrt{2}}\{1 + s + [(1 + s)^2 + (t - \gamma)^2]^{1/2}\}^{1/2}, \\ k_1 &= \frac{1}{\sqrt{2}}\{-(1 + s) + [(1 + s)^2 + (t - \gamma)^2]^{1/2}\}^{1/2}, \\ j_2 &= \frac{1}{\sqrt{2}}\{(1 - s) + [(1 - s)^2 + (t + \gamma)^2]^{1/2}\}^{1/2}, \\ k_2 &= \frac{1}{\sqrt{2}}\{-(1 - s) + [(1 - s)^2 + (t + \gamma)^2]^{1/2}\}^{1/2}. \end{aligned} \tag{3.65}$$

The eight roots of the auxiliary equation are then given by

$$\begin{aligned} &j_1 \pm ik_1, \qquad -(j_1 \pm ik_1), \\ &j_2 \pm ik_2, \qquad -(j_2 \pm ik_2). \end{aligned}$$

It follows from the remarks in connection with (3.63) that r_1 and r_2 correspond to the upper and the lower sign, respectively, in equation (3.55). We conclude that

$$\begin{aligned}(r_1^2 - 1)^2 &= 2i\gamma(r_1^2 - k),\\ (r_2^2 - 1)^2 &= -2i\gamma(r_2^2 - k).\end{aligned} \tag{3.66}$$

3.3.2 *Solution of the homogeneous equation*

Corresponding to the eight roots of the auxiliary equation, we have the complex solutions

$$e^{r_1\eta}, \qquad e^{r_2\eta}, \qquad \ldots, \qquad e^{r_8\eta}.$$

We now consider two conjugate complex roots, for instance, r_1 and $\bar{r}_1$. For the corresponding exponential functions, on noting the relation

$$e^{(x+iy)} = e^x(\cos y + i \sin y),$$

we obtain

$$\left.\begin{matrix} e^{r_1\eta} \\ e^{\bar{r}_1\eta} \end{matrix}\right\} = e^{-j_1\eta}\cos k_1\eta \pm e^{-j_1\eta}\sin k_1\eta = f_1(\eta) \pm i g_1(\eta). \tag{3.67}$$

It is shown in the theory of linear differential equations that the above functions $f_1(\eta)$ and $g_1(\eta)$ are real solutions of the homogeneous differential equation (3.52), so that, altogether, we obtain eight real, linearly independent solutions. An arbitrary linear combination $(a_1 f_1 + b_1 g_1)$ is likewise a solution and may be written in the form

$$a_1 f_1 + b_1 g_1 = \mathrm{Re}[(a_1 - ib_1)(f_1 + ig_1)] = \mathrm{Re}(A_1\, e^{r_1\eta}),$$

where $A_1 = (a_1 - ib_1)$ is a complex constant.

If we apply a linear differential operator with constant, real coefficients p_ν, $\nu = 0, 1, 2, \ldots, n$, to the function $(a_1 f_1 + b_1 g_1)$, we obtain

$$\begin{aligned} N[a_1 f_1 + b_1 g_1] &= \sum_{\nu=0}^{n} p_\nu \frac{d^\nu}{d\eta^\nu}(a_1 f_1 + b_1 g_1) \\ &= \sum_{\nu=0}^{n} p_\nu \frac{d^\nu}{d\eta^\nu}\mathrm{Re}(A_1 e^{r_1\eta}) \\ &= \sum_{\nu=0}^{n} \mathrm{Re}\left(p_\nu \frac{d^\nu}{d\eta^\nu}(A_1 e^{r_1\eta})\right) \\ &= \mathrm{Re}\left(\sum_{\nu=0}^{n} (p_\nu r_1^{\nu}) A_1 e^{r_1\eta}\right) \\ &= \mathrm{Re}[P(r_1) A_1 e^{r_1\eta}], \end{aligned} \tag{3.68}$$

where

$$P(r_1) = \sum_{\nu=0}^{n} (p_\nu r_1^\nu) \tag{3.68a}$$

is a polynomial in the root of the auxiliary equation (the *characteristic polynomial*).

Let

$$P(r_1) = R_1 + iJ_1,$$

then we have

$$\begin{aligned} \mathrm{Re}[P(r_1)A_1 e^{r_1\eta}] &= \mathrm{Re}[(R_1 + iJ_1)(a_1 - ib_1)(f_1 + ig_1)] \\ &= R_1(a_1 f_1 + b_1 g_1) + J_1(b_1 f_1 - a_1 g_1) \\ &= [R_1 \quad J_1]\begin{bmatrix} f_1 & g_1 \\ -g_1 & f_1 \end{bmatrix}\begin{bmatrix} a_1 \\ b_1 \end{bmatrix}, \end{aligned} \tag{3.69}$$

and similar expressions hold for the other roots.

The complete solution of the homogeneous equation (3.52) can now be written

$$\tilde{\psi}^h(\eta) = \mathrm{Re}(A_1 e^{r_1\eta} + A_2 e^{r_2\eta} + A_3 e^{-r_1\eta} + A_4 e^{-r_2\eta}), \tag{3.70}$$

since the two real solutions corresponding to the roots r_1 and $\bar{r}_1$ are represented by the term $\mathrm{Re}(A_1 e^{r_1\eta})$. The quantities A_1, A_2, A_3, and A_4 in (3.70) are arbitrary complex constants.

In the first two terms of (3.70), we shall now introduce a new variable ζ instead of η, where ζ is proportional to the distance from the left-hand edge $y = -b/2$. Similarly, in the last two terms, we shall introduce a variable θ proportional to the distance from the right-hand edge $y = b/2$ (see Fig. 34), where

$$\begin{aligned} \zeta &= \eta + \tfrac{1}{2}\alpha_n b = \alpha_n(y + b/2), \\ \theta &= \tfrac{1}{2}\alpha_n b - \eta = \alpha_n(b/2 - y). \end{aligned}$$

We thus obtain

$$\begin{aligned} &\mathrm{Re}\{A_1 \exp[r_1(\zeta - \tfrac{1}{2}\alpha_n b)] + \ldots + A_3 \exp[-r_1(\tfrac{1}{2}\alpha_n b - \theta)] + \ldots\} \\ &\qquad = \mathrm{Re}\{[A_1 \exp(-\tfrac{1}{2}r_1\alpha_n b)]e^{r_1\zeta} + \ldots + [A_3 \exp(-\tfrac{1}{2}r_1\alpha_n b)]e^{r_1\theta} + \ldots\}. \end{aligned}$$

It will now be seen that the coefficients of $e^{r_1\zeta}$ and $e^{r_1\theta}$ in the last line of the formula may be regarded as modified arbitrary constants. The complete

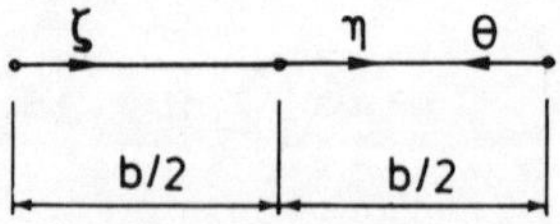

Fig. 34. Distances from left-hand and right-hand edges

solution of the homogeneous differential equation (3.52) may therefore be written in the form

$$\tilde{\psi}^{\mathrm{h}} = \mathrm{Re}(A_1 \mathrm{e}^{r_1\zeta} + A_2 \mathrm{e}^{r_2\zeta} + A_3 \mathrm{e}^{r_1\theta} + A_4 \mathrm{e}^{r_2\theta}), \tag{3.71}$$

where A_1, A_2, A_3, and A_4 now denote the modified arbitrary constants. The first two terms on the right-hand side represent *edge effects from the left-hand edge*, and they are damped out with increasing ζ, since

$$\mathrm{e}^{r_1\zeta} = \mathrm{e}^{-j_1\zeta}\,\mathrm{e}^{\mathrm{i}k_1\zeta}, \qquad \text{and} \qquad j_1 > 0$$

(see $(3.65)_1$). Similarly, the last two terms of (3.71) represent *edge effects from the right-hand edge*.

So far as the solution of the homogeneous equations is concerned, it follows from (3.34) and (3.35) that the normal displacement w must satisfy the same homogeneous, partial differential equation as the stress function ψ. Introducing the expression $(3.38)_2$ for w and the dimensionless function $(3.46)_3$, it will be seen that $\tilde{w}^{\mathrm{h}}$ satisfies a differential equation of the same form as the equation (3.52) for $\tilde{\psi}^{\mathrm{h}}$. The roots of the auxiliary equation are therefore the same as before, so that the complete solution for $\tilde{w}^{\mathrm{h}}$ can be written in a form similar to (3.71), i.e.

$$\tilde{w}^{\mathrm{h}} = \mathrm{Re}(B_1 \mathrm{e}^{r_1\zeta} + B_2 \mathrm{e}^{r_2\zeta} + B_3 \mathrm{e}^{r_1\theta} + B_4 \mathrm{e}^{r_2\theta}). \tag{3.72}$$

We shall now investigate *the connection between $\tilde{w}^{\mathrm{h}}$ and $\tilde{\psi}^{\mathrm{h}}$*. For this purpose we shall use equation (3.30). Inserting (3.38) into (3.30), we get

$$\left(\frac{\mathrm{d}^2}{\mathrm{d}y^2} - \alpha_n^2\right)^2 \psi_n^{\mathrm{h}} + Eh\left(\frac{1}{R_x}\frac{\mathrm{d}^2}{\mathrm{d}y^2} - \frac{\alpha_n^2}{R_y}\right) w_n^{\mathrm{h}} = 0,$$

which, by means of (3.46) and (3.52a), can be reduced to the dimensionless form

$$\left(\frac{\mathrm{d}^2}{\mathrm{d}\eta^2} - 1\right)^2 \tilde{\psi}^{\mathrm{h}} + 2\gamma\left(\frac{\mathrm{d}^2}{\mathrm{d}\eta^2} - k\right)\tilde{w}^{\mathrm{h}} = 0. \tag{3.73}$$

If we substitute the expressions (3.71) and (3.72) for $\tilde{\psi}^{\mathrm{h}}$ and $\tilde{w}^{\mathrm{h}}$, we obtain an equation of the form

$$\mathrm{Re}(H_1 \mathrm{e}^{r_1\zeta} + H_2 \mathrm{e}^{r_2\zeta} + H_3 \mathrm{e}^{r_1\theta} + H_4 \mathrm{e}^{r_2\theta}) = 0. \tag{3.74}$$

Now, we have the relations

$$\mathrm{Re}(H_1 \mathrm{e}^{r_1\zeta}) = \mathrm{Re}[(h_1 - \mathrm{i}l_1)(f_1(\zeta) + \mathrm{i}g_1(\zeta))] = h_1 f_1(\zeta) + l_1 g_1(\zeta),$$

and similar expressions hold for the other terms. As the eight real solutions of the homogeneous equation are linearly independent, it follows that the left-hand side of (3.74) can only vanish when $h_1 = l_1 = \ldots = 0$, i.e. when

$$H_1 = H_2 = H_3 = H_4 = 0.$$

The resulting complex constants should therefore be zero.

We first consider the edge effects from the left-hand edge. Noting that

$$\frac{d}{d\eta}(\ ) = \frac{d}{d\zeta}(\ ),$$

we get, corresponding to the terms containing $e^{r_1\zeta}$,

$$H_1 = (r_1^2 - 1)^2 A_1 + 2\gamma(r_1^2 - k)B_1 = 0,$$

or, by using $(3.66)_1$,

$$B_1 = -iA_1. \tag{3.75a}$$

In a similar manner we find, corresponding to the root r_2,

$$B_2 = iA_2. \tag{3.75b}$$

The representation of v_1 *and* v_2. Before continuing the argument of the present section, we shall make a short digression in order to prove that the representation (3.45) used for the displacements v_1 and v_2 is, in fact, permissible. We assume that we have determined a solution of the type (3.38), and corresponding functions v_1 and v_2 of the type (3.45), so that the equations $(3.36)_{11,12}$ for $\partial v_1/\partial x$ and $\partial^2 v_2/\partial x^2$ are satisfied.

Our assertion that the functions v_1 and v_2 are, in fact, the tangential displacements will be proved if we can show that the strains determined by means of v_1 and v_2 are identical with the strains produced by the contact forces, i.e. that the following equations are satisfied:

$$\frac{\partial v_1}{\partial x} - \frac{w}{R_x} = \frac{1}{Eh}\psi_{,yy}, \qquad \frac{\partial v_2}{\partial y} - \frac{w}{R_y} = \frac{1}{Eh}\psi_{,xx},$$

$$\frac{1}{2}\left(\frac{\partial v_1}{\partial y} + \frac{\partial v_2}{\partial x}\right) = -\frac{1}{Eh}\psi_{,xy}.$$

If we introduce dimensionless functions (see Table 3.1), these equations assume the form

$$\tilde{v}_1 - 2\gamma\tilde{w} = \tilde{\psi}'', \qquad \tilde{v}_2' - 2\gamma k\tilde{w} = -\tilde{\psi}, \qquad \tilde{v}_1' - \tilde{v}_2 = 2\tilde{\psi}'. \tag{3.76}$$

It will now be seen that the first and third equations are satisfied because of $(3.49)_{10,11}$, which are fulfilled by assumption. In order to prove the second equation $(3.76)_2$, we differentiate $(3.49)_{11}$ to obtain

$$\tilde{v}_2' = \tilde{\psi}'''' - 2\tilde{\psi}'' + 2\gamma\tilde{w}'', \tag{3.77}$$

and we use the governing equation (3.30), which, by (3.73), assumes the dimensionless form

$$\tilde{\psi}'''' - 2\tilde{\psi}'' + \tilde{\psi} + 2\gamma\tilde{w}'' - 2\gamma k\tilde{w} = 0. \tag{3.78}$$

On substituting from (3.77) into (3.78), we derive the second of equations (3.76). This completes the proof.

In section 3.3 we derived the equations (3.49), which express the contact forces and couples and the displacements in terms of $\tilde{\psi}$ and $\tilde{w}$. We shall now determine the contributions to these statical and geometrical quantities from the solution of the homogeneous equations.

It will be seen that the right-hand sides of (3.49) consist of linear differential operators with constant coefficients applied to $\tilde{\psi}$ and $\tilde{w}$. Corresponding to those terms of $\tilde{\psi}^{\mathrm{h}}$ and $\tilde{w}^{\mathrm{h}}$ that contain the same exponential function $e^{r\zeta}$, we obtain a contribution which, according to (3.68), can be written in the form

$$\mathrm{Re}[P(r)A\,e^{r\zeta}],$$

where the connection (3.75) between the arbitrary constants for $\tilde{w}^{\mathrm{h}}$ and $\tilde{\psi}^{\mathrm{h}}$ is used, and $P(r)$ is a characteristic polynomial.

The contribution to one of the quantities (3.49) from the edge effects from the left-hand edge can now be written

$$\mathrm{Re}[P(r_1)A_1e^{r_1\zeta} + P(r_2)A_2e^{r_2\zeta}]$$

$$= [R_1 \quad J_1 \quad R_2 \quad J_2]\begin{bmatrix} f_1(\zeta) & g_1(\zeta) & 0 & 0 \\ -g_1(\zeta) & f_1(\zeta) & 0 & 0 \\ 0 & 0 & f_2(\zeta) & g_2(\zeta) \\ 0 & 0 & -g_2(\zeta) & f_2(\zeta) \end{bmatrix}\begin{bmatrix} a_1 \\ b_1 \\ a_2 \\ b_2 \end{bmatrix}, \qquad (3.79)$$

where (3.69) has been used, and the quantities appearing in the last matrix expression are defined by

$$P(r_1) = R_1 + iJ_1, \qquad P(r_2) = R_2 + iJ_2,$$
$$e^{r_1\zeta} = f_1 + ig_1, \qquad e^{r_2\zeta} = f_2 + ig_2,$$
$$A_1 = a_1 - ib_1, \qquad A_2 = a_2 - ib_2 \qquad \text{(arbitrary constants)}.$$

By using the relations (3.75), so that

$$\tilde{\psi}^{\mathrm{h}} \sim \mathrm{Re}(A\,e^{r\zeta}), \qquad \tilde{w}^{\mathrm{h}} \sim \mathrm{Re}(\mp iA\,e^{r\zeta}),$$

we obtain the following characteristic polynomials from equations (3.49):

$$\tilde{N}_x : r^2, \qquad \tilde{N}_{xy} : r, \qquad \tilde{N}_y : -1,$$
$$\tilde{M}_x : \mp i, \qquad \tilde{M}_{xy} : \mp ir, \qquad \tilde{M}_y : \pm ir^2,$$
$$\tilde{Q}_x : \mp i(r^2 - 1), \qquad \tilde{Q}_y : \pm ir(r^2 - 1),$$
$$\tilde{S}_y : \pm ir(r^2 - 2), \qquad \tilde{w} : \mp i,$$
$$\tilde{v}_1 : r^2 \mp 2i\gamma, \qquad \tilde{v}_2 : r(r^2 - 2) \mp 2i\gamma r,$$
$$\tilde{\omega}_2 : \pm ir.$$

In these equations, the upper signs apply to the root r_1, and the lower signs to the root r_2. The roots and their squares are given by (3.64) and (3.63).

By using these formulae, we also find the relations

$$r_1(r_1^2-2)=j_1(1-s)+k_1(t-\gamma)+\mathrm{i}[j_1(t-\gamma)-k_1(1-s)],$$
$$r_2(r_2^2-2)=j_2(1+s)+k_2(t+\gamma)+\mathrm{i}[j_2(t+\gamma)-k_2(1+s)].$$

The real and imaginary parts of the characteristic polynomials can now be calculated. The resulting expressions are tabulated in Table 3.2.

The statical and geometrical quantities are now assembled in a column vector **f**, where

$$\mathbf{f}^{\mathrm{T}}=[\tilde{N}_{xy}\quad \tilde{S}_y\quad \tilde{N}_y\quad \tilde{M}_y\quad \tilde{v}_1\quad \tilde{w}\quad \tilde{v}_2\quad \tilde{\omega}_2\quad \tilde{N}_x\quad \tilde{M}_x\quad \tilde{M}_{xy}\quad \tilde{Q}_x\quad \tilde{Q}_y]. \tag{3.80}$$

The first eight components are the statical and geometrical quantities that may be required for the purpose of formulating the edge conditions at the edges $y=\pm b/2$ (see also Fig. 35).

Using (3.79), the contribution from the edge effects from the left-hand edge corresponding to $\tilde{\psi}^{\mathrm{h}}$ and $\tilde{w}^{\mathrm{h}}$ can be written in matrix form as follows:

$$\mathbf{f}=\mathbf{AF}(\zeta)\mathbf{a}, \tag{3.81}$$

where

$$\mathbf{F}(\zeta)=\begin{bmatrix} f_1(\zeta) & g_1(\zeta) & 0 & 0\\ -g_1(\zeta) & f_1(\zeta) & 0 & 0\\ 0 & 0 & f_2(\zeta) & g_2(\zeta)\\ 0 & 0 & -g_2(\zeta) & f_2(\zeta)\end{bmatrix},\qquad \mathbf{a}=\begin{bmatrix} a_1\\ b_1\\ a_2\\ b_2\end{bmatrix}, \tag{3.81a}$$

and the matrix **A** is shown in Table 3.2. This completes the treatment of the edge effects from the left-hand edge.

The edge effects from the right-hand edge correspond to the terms

$$\tilde{\psi}^{\mathrm{h}}\sim \mathrm{Re}(A_3\mathrm{e}^{r_1\theta}+A_4\mathrm{e}^{r_2\theta}),\qquad \text{where}\qquad \theta=\tfrac{1}{2}\alpha_n b-\eta.$$

We note the following relation between the derivatives with respect to η and θ:

$$\frac{\mathrm{d}}{\mathrm{d}\eta}(\ )=-\frac{\mathrm{d}}{\mathrm{d}\theta}(\ ). \tag{3.82}$$

In the present case, the connection between the arbitrary constants A_3 and A_4 associated with $\tilde{\psi}^{\mathrm{h}}$ and the constants B_3 and B_4 associated with $\tilde{w}^{\mathrm{h}}$ is given

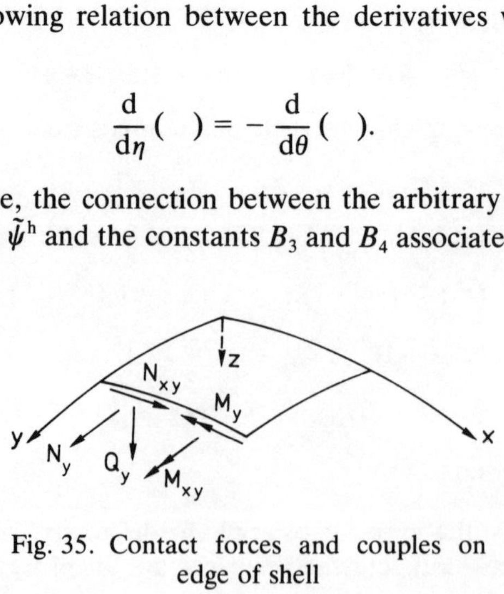

Fig. 35. Contact forces and couples on edge of shell

Table 3.2 Expressions for the real and imaginary parts of the characteristic polynomials

	J	R_1	J_1	R_2	J_2
$\tilde{N}_{xy}$	−1	$-j_1$	k_1	$-j_2$	k_2
$\tilde{S}_y$	−1	$-j_1(t-\gamma)+k_1(1-s)$	$j_1(1-s)+k_1(t-\gamma)$	$j_2(t+\gamma)-k_2(1+s)$	$-j_2(1+s)-k_2(t+\gamma)$
$\tilde{N}_y$	1	-1	0	-1	0
$\tilde{M}_y$	1	$t-\gamma$	$1+s$	$-(t+\gamma)$	$-(1-s)$
$\tilde{v}_1$	1	$1+s$	$-(t+\gamma)$	$1-s$	$-(t-\gamma)$
$\tilde{w}$	1	0	-1	0	1
$\tilde{v}_2$	−1	$j_1(1-s)+k_1(t+\gamma)$	$j_1(t+\gamma)-k_1(1-s)$	$j_2(1+s)+k_2(t-\gamma)$	$j_2(t-\gamma)-k_2(1+s)$
$\tilde{\omega}_2$	−1	$-k_1$	$-j_1$	k_2	j_2
$\tilde{N}_x$	1	$1+s$	$-(t-\gamma)$	$1-s$	$-(t+\gamma)$
$\tilde{M}_x$	1	0	-1	0	1
$\tilde{M}_{xy}$	−1	k_1	j_1	$-k_2$	$-j_2$
$\tilde{Q}_x$	1	$-(t-\gamma)$	$-s$	$t+\gamma$	$-s$
$\tilde{Q}_y$	−1	$-j_1(t-\gamma)-k_1s$	$-j_1s+k_1(t-\gamma)$	$j_2(t+\gamma)-k_2s$	$-j_2s-k_2(t+\gamma)$

Matrix **A**

by the formulae

$$B_3 = -iA_3, \qquad B_4 = iA_4, \tag{3.83}$$

which have exactly the same form as (3.75), since only derivatives of even order with respect to η appear in the equations used to derive (3.83).

It will be seen that each of the right-hand sides of (3.49) contains either derivatives of even order with respect to η, or derivatives of odd order. It follows (because of (3.82)) that the characteristic polynomials will be either the same as before (for derivatives of even order) or of opposite sign (for derivatives of odd order). This means that a change of signs occurs for those quantities in Table 3.2 for which the number -1 appears in the second column.

The contribution from the edge effects from the right-hand edge can now be written in the form

$$\mathbf{f} = \mathbf{JAF}(\theta)\mathbf{b}, \tag{3.84}$$

where $\mathbf{J}$ is a diagonal matrix, the diagonal elements of which are given by the second column in Table 3.2, while the vector

$$\mathbf{b}^{\mathrm{T}} = [a_3 \quad b_3 \quad a_4 \quad b_4] \tag{3.84a}$$

contains the arbitrary constants associated with the edge effects from the right-hand edge.

3.3.3 Particular integral

We shall here confine ourselves to the simple case in which the load p depends only on x. The loading term (3.37a) then assumes the form

$$p(x) = p_n \cos \alpha_n x, \tag{3.85a}$$

where p_n is a constant. This is written

$$p(x) = p_0 \tilde{p} \cos \alpha_n x, \tag{3.85b}$$

where

$$\tilde{p} = p_n/p_0 \tag{3.85c}$$

is a dimensionless constant, and we seek a corresponding particular integral of the form

$$\begin{aligned} \psi^{\mathrm{P}}(x) &= \frac{p_0 R_y}{\alpha_n^2} \tilde{\psi}^{\mathrm{P}} \cos \alpha_n x, \\ w^{\mathrm{P}}(x) &= \frac{p_0 R_y}{Eh^2} \frac{2\sqrt{3}}{\alpha_n^2} \tilde{w}^{\mathrm{P}} \cos \alpha_n x, \end{aligned} \tag{3.86}$$

where $\tilde{\psi}^{\mathrm{P}}$ and $\tilde{w}^{\mathrm{P}}$ are dimensionless constants. Substituting the expressions (3.85b) and $(3.86)_1$ into the differential equation (3.34), and noting that all derivatives with respect to y vanish, we obtain

$$\left(\alpha_n^8 + \frac{12}{h^2}\frac{\alpha_n^4}{R_y^2}\right)\frac{p_0 R_y}{\alpha_n^2}\tilde{\psi}^{\mathrm{p}} = \frac{12}{h^2}\frac{\alpha_n^2}{R_y}p_0\tilde{p}.$$

With the help of (3.52a), this reduces to

$$(1 + 4\gamma^2 k^2)\tilde{\psi}^{\mathrm{p}} = 4\gamma^2 k^2 \tilde{p},$$

or

$$\tilde{\psi}^{\mathrm{p}} = \frac{4\gamma^2 k^2}{1 + 4\gamma^2 k^2}\tilde{p}. \tag{3.87a}$$

Using (3.30), we further derive

$$\tilde{\psi}^{\mathrm{p}} - 2\gamma k \tilde{w}^{\mathrm{p}} = 0,$$

or

$$w^{\mathrm{p}} = \tilde{\psi}^{\mathrm{p}}/(2\gamma k). \tag{3.87b}$$

The corresponding contact forces and couples and the displacements can now be determined by means of equations (3.49). Inserting the expressions (3.87a) and (3.87b) into these equations, we get the results tabulated in Table 3.3. These quantities are assembled into a column vector $\mathbf{f}^{\mathrm{p}}$ (cf. (3.80)).

Table 3.3 Particular integral

$\tilde{N}_{xy}$	$\tilde{S}_y$	$\tilde{N}_y$	$\tilde{M}_y$	$\tilde{v}_1$	$\tilde{w}$	$\tilde{v}_2$
0	0	$-\tilde{\psi}^{\mathrm{p}}$	0	$2\gamma\tilde{w}^{\mathrm{p}}$	$\tilde{w}^{\mathrm{p}}$	0
$\tilde{\omega}_2$	$\tilde{N}_x$	$\tilde{M}_x$	$\tilde{M}_{xy}$	$\tilde{Q}_x$	$\tilde{Q}_y$	
0	0	$\tilde{w}^{\mathrm{p}}$	0	$-\tilde{w}^{\mathrm{p}}$	0	

3.3.4 *Edge conditions*

The complete solution can now be written as the sum of the particular integral and the solution of the homogeneous equations. From (3.81) and (3.84) we obtain the following equation in terms of the dimensionless functions

$$\mathbf{f} = \mathbf{f}^{\mathrm{p}} + \mathbf{AF}(\zeta)\mathbf{a} + \mathbf{JAF}(\theta)\mathbf{b}. \tag{3.88}$$

This expression can now be used to formulate the edge conditions for the edges $y = \pm b/2$.

In the case of a shallow shell, the three types of edge conditions treated in section 1.7 are expressed in the following manner (assuming that the boundary curve is part of a θ_1-curve, so that y = constant on the edge):

(a) *Clamped edge*

$$v_1 = 0, \quad v_2 = 0, \quad w = 0, \quad \frac{\partial w}{\partial y} = 0. \tag{3.89}$$

(b) *Simply supported edge*

$$v_1 = 0, \qquad v_2 = 0, \qquad w = 0, \qquad M_y = 0. \tag{3.90}$$

(c) *Free edge*

$$\begin{aligned} &N_y = 0, \qquad N_{xy} = 0, \qquad M_y = 0, \\ &S_y = Q_y + M_{xy,x} = 0. \end{aligned} \tag{3.91}$$

In the case of a shallow shell over a rectangular plane, we introduce dimensionless functions according to (3.45) and Table 3.1, and it will be seen that we may then replace the statical and geometrical quantities in the above edge conditions (3.89), (3.90), and (3.91) by the corresponding dimensionless functions, and replace $\partial(\ \ )/\partial y$ by $\partial(\ \ )/\partial \eta$.

(d) *Edge beam supported by columns*. We shall now consider a fourth type of edge condition. Let us assume that an elastic edge beam supported by fairly closely spaced vertical columns is provided at the edge $y = -b/2$. We shall derive a set of edge conditions that furnishes an approximate description of this type of support. The derivation will be based on the following approximations.

(1) The edge beam is regarded as a perfectly flexible member so that the internal moments are zero and the only internal force is the direct force N.

(2) For the purpose of formulating the edge conditions, we shall replace the vertical columns by columns whose axes coincide with the normals to the middle surface at the edge.

(3) The columns are assumed to be inextensible bars connected to the rest of the structure by frictionless hinges.

It follows from these assumptions that the following conditions are satisfied at the edge:

$$w = 0, \qquad N_y = 0, \qquad M_y = 0. \tag{3.92a}$$

The fourth edge condition is derived in the following manner. Since $w = 0$ for $y = -b/2$, we conclude that the strain of the middle surface along the edge is given by $\varepsilon_{11} = \partial v_1/\partial x$ (see (3.9)), and we note that this strain equals the extensional strain of the edge beam. Moreover, we have the following equation of equilibrium for the beam (see Fig. 36)

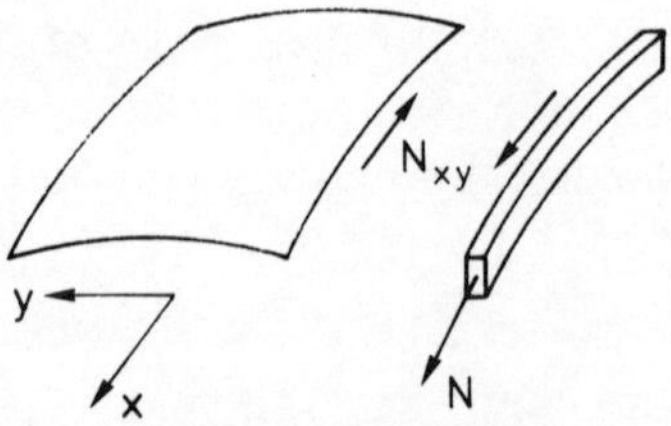

Fig. 36. Forces on edge beam

$$\frac{dN}{dx} + N_{xy} = 0,$$

together with the constitutive equation of the beam, i.e.

$$N = E_b A \varepsilon_{11} = E_b A \frac{\partial v_1}{\partial x},$$

where E_b is Young's modulus and A is the cross-sectional area of the edge beam. From these equations we deduce the condition

$$-E_b A \frac{\partial^2 v_1}{\partial x^2} = N_{xy} \qquad \text{for} \qquad y = -b/2. \tag{3.92b}$$

The four edge conditions for the type (d) support are therefore given approximately by the equations (3.92a and b).

Introducing dimensionless functions for the case of a shallow shell over a rectangular plan, we find that the edge conditions (3.92) assume the form

$$\left.\begin{aligned} &\tilde{w} = \tilde{N}_y = \tilde{M}_y = 0, \\ &\tilde{N}_{xy} - \frac{E_b A \alpha_n}{Eh} \tilde{v}_1 = 0, \end{aligned}\right\} \qquad \text{for} \qquad \eta = -\tfrac{1}{2}\alpha_n b. \tag{3.93}$$

We now consider a simple example of the application of the edge conditions. Let us assume that the edge $y = -b/2$ is free, while the edge $y = b/2$ is clamped. This means that the first four components of the vector **f** should be zero at the left-hand edge (where $\zeta = 0$ and $\theta = \alpha_n b$) while the next four components of **f** should be zero at the right-hand edge (where $\theta = 0$ and $\zeta = \alpha_n b$). According to (3.88), the corresponding edge conditions are expressed by the equations

$$\begin{aligned} &[\mathbf{f}^p + \mathbf{A}\mathbf{a} + \mathbf{J}\mathbf{A}\mathbf{F}(\alpha_n b)\mathbf{b}]_{1-4} = \mathbf{0}, \\ &[\mathbf{f}^p + \mathbf{A}\mathbf{F}(\alpha_n b)\mathbf{a} + \mathbf{J}\mathbf{A}\mathbf{b}]_{5-8} = \mathbf{0}, \end{aligned} \tag{3.94}$$

where indices 1–4 and 5–8 denote elements one to four and elements five to eight, respectively, of the corresponding column vectors. Note that $\mathbf{F}(0) = \mathbf{I}$ (the (4, 4) unit matrix) (see (3.81a) and (3.67)). Equations (3.94) constitute a system of eight linear equations for the eight arbitrary constants.

3.4 The Donnell theory for circular cylindrical shells

It will be recalled that two types of simplifications were introduced in the derivation of the theory of shallow shells (see section 3.1).

Simplification 1 implied that the contributions from the tangential displacements v_1 and v_2 to the bending measures, as well as the contributions from the transverse shear forces Q_1 and Q_2 to the first two equations of equilibrium (1.111), could be neglected.

Simplification 2 was based on the assumption that the shell is shallow, and it implied that the coefficients of the first and second fundamental forms and the Lamé parameters of the middle surface are given approximately by

$$a_{\alpha\beta} = \delta_{\alpha\beta}, \qquad A_1 = A_2 = 1, \qquad b_{\alpha\beta} = z_{,\alpha\beta}. \tag{3.95}$$

We now consider a *circular cylindrical shell* with generators parallel to the X-axis. We shall use the distance x measured along the generator and the arc length s of the circular cross-section as curvilinear coordinates on the middle surface, so that (see Fig. 37)

$$\theta_1 = x, \qquad \theta_2 = s.$$

For these curvilinear coordinates, the following results are easily established:

$$a_{\alpha\beta} = \delta_{\alpha\beta}, \qquad A_1 = A_2 = 1,$$
$$[b_{\alpha\beta}] = \begin{bmatrix} 0 & 0 \\ 0 & 1/R \end{bmatrix}, \qquad \frac{1}{R_1} = 0, \qquad \frac{1}{R_2} = \frac{1}{R}. \tag{3.96}$$

It will now be seen that the first two equations (3.95), which were approximately true for the shallow shell, are satisfied exactly for the cylindrical shell (developable surface), and that the coordinate curves are the lines of curvature of the cylinder, since $a_{12} = 0$ and $b_{12} = 0$.

We shall now derive a set of governing equations for the circular cylindrical shell based on the above *simplification 1*. In this way we arrive at the so-called *Donnell theory* for cylindrical shells.

Using (3.96), we find for the strain measures (1.66):

$$\varepsilon_{11} = v_{1,1}, \qquad \varepsilon_{22} = v_{2,2} - (w/R), \qquad \varepsilon_{12} = \tfrac{1}{2}(v_{1,2} + v_{2,1}).$$

It will be seen that these expressions together with the approximate expressions (3.1) for the bending measures, the approximate equations of equilibrium (3.2), and the constitutive equations (1.135), in the present case of a cylindrical shell become identical to the corresponding equations (3.9), (3.10), and (3.18) for a shallow shell whose middle surface is given by (3.28), and for which

$$1/R_x = 0, \qquad 1/R_y = 1/R.$$

The governing equations of the cylindrical shell are therefore the same as

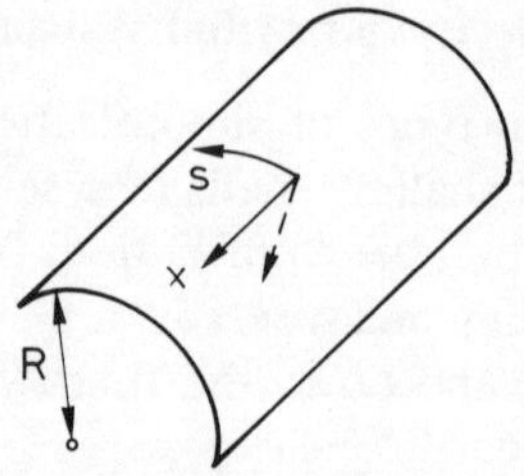

Fig. 37. Circular cylindrical shell

the governing equations of the shallow shell. For the cylindrical shell, we therefore have (see (3.32))

$$L[\ \ ] = \frac{1}{R}\frac{\partial^2}{\partial x^2}(\ \ ),$$

and the equations (3.30), (3.31), and (3.34) assume the form

$$\nabla^4\psi + \frac{Eh}{R}\frac{\partial^2 w}{\partial x^2} = 0, \qquad \frac{Eh^3}{12}\nabla^4 w - \frac{1}{R}\frac{\partial^2\psi}{\partial x^2} = p,$$
$$\nabla^8\psi + \frac{12}{h^2R^2}\frac{\partial^4\psi}{\partial x^4} = -\frac{12}{h^2R}\frac{\partial^2 p}{\partial x^2}. \tag{3.97}$$

It should be noted that the *analytical solution* for a shallow shell over a rectangular plan derived in section 3.3 can be used without change for a circular cylindrical shell, the edges of which are given by $x = \pm l/2, s = \pm b/2$. The dimensionless parameters introduced in section 3.3.1 in connection with the solution of the auxiliary equation are now given by

$$\gamma = 0, \qquad \beta = \frac{2\surd 3}{h\alpha_n^2 R}, \qquad s = t = \surd(\beta/2), \tag{3.98}$$

while the quantity γk appearing in the particular integral assumes the value (see (3.52a))

$$\gamma k = \frac{\surd 3}{h\alpha_n^2 R} = \tfrac{1}{2}\beta \tag{3.98a}$$

for the cylindrical shell. Finally, it should be emphasized that the present method of analysis can be used for non-shallow cylindrical shells, since it was not assumed in the previous derivations that the cylindrical shell is shallow.

3.5 Shallow spherical shell

We consider a shallow spherical shell of constant thickness, and we restrict ourselves to a treatment of the axisymmetric case. This problem can be solved by means of the above theory of shallow shells if we introduce a polar coordinate system in the *XY*-plane. However, we shall here proceed in a somewhat different manner. We apply the theory of shells of revolution developed in Chapter 2, together with appropriate simplifications which express the fact that the shell is shallow.

We assume that the condition

$$\varphi^2 \ll 1 \tag{3.99}$$

is satisfied everywhere on the middle surface (the meaning of the angle φ is shown in Fig. 19, section 2.2). This condition corresponds to (3.4) and implies that the angle φ is a small quantity, i.e. that the shell is shallow. In the series expansions for the trigonometric functions, we shall therefore neglect terms of the second and higher degree in φ (cf. simplification 2 in section 3.1), i.e. we shall use the approximations

$$\sin\varphi \simeq \varphi, \qquad \cos\varphi \simeq 1. \tag{3.100a}$$

For the function $\cot\varphi$ we have the following series expansion (Laurent series)

$$\cot\varphi = \frac{1}{\varphi} - \left(\frac{\varphi}{3} + \frac{\varphi^3}{45} + \ldots\right) = \frac{1}{\varphi}\left(1 - \frac{\varphi^2}{3} - \frac{\varphi^4}{45} - \ldots\right), \tag{3.101}$$

which is valid for $0 < |\varphi| < \pi$. We therefore introduce the approximation

$$\cot\varphi \simeq 1/\varphi. \tag{3.100b}$$

In the present case of a shallow shell, we shall also omit the contributions from the tangential displacement u to the bending measures, as well as the contribution from the transverse shear force Q_1 to the first equation of equilibrium (2.10) (cf. simplification 1 in section 3.1). When the governing equations of section 2.2 are modified in this manner, and the approximations (3.100a) and (3.100b) are introduced, the following expressions are obtained (note that $R_1 = R_2 = R$ for the spherical shell):

Strain and bending measures. Equations (2.7) and (2.8) assume the form

$$\begin{aligned} \varepsilon_{11} &= \frac{1}{R}(u' - w), & \varepsilon_{22} &= \frac{1}{R}\left(\frac{u}{\varphi} - w\right), \\ \kappa_{11} &= \frac{-w''}{R^2}, & \kappa_{22} &= -\frac{1}{R^2\varphi}w'. \end{aligned} \tag{3.102}$$

Compatibility condition. Equation (2.19) assumes the form

$$\varepsilon_{22}' + \frac{1}{\varphi}(\varepsilon_{22} - \varepsilon_{11}) = -\chi. \tag{3.103}$$

Equations of equilibrium. Since we have the relation $r = R\varphi$ for the shallow shell (cf. $(2.1)_2$), the equations (2.10) assume the form

$$\begin{aligned} N_{11}' + (1/\varphi)(N_{11} - N_{22}) + Rp_1 &= 0, \\ Q_1' + (1/\varphi)Q_1 + N_{11} + N_{22} + Rp_3 &= 0, \\ M_{11}' + (1/\varphi)(M_{11} - M_{22}) - RQ_1 &= 0. \end{aligned} \tag{3.104}$$

The constitutive equations are given by (2.11).

We first consider the solution of the homogeneous equations ($p_1 = p_3 = 0$). For the shallow spherical shell, the equations (2.12) and (2.13) assume the form

$$\begin{aligned} \chi &= w'/R, & \psi &= -RQ_1, \\ \kappa_{11} &= -\chi'/R, & \kappa_{22} &= -(1/R\varphi)\chi. \end{aligned} \tag{3.105}$$

If the bending moments $(2.11)_{3,4}$ are expressed in terms of the rotation χ by means of (3.105), and the resulting formulae for the moments are inserted in the third equation of equilibrium (3.104), we obtain the following equation (which corresponds to $(2.25)_1$):

$$L_s[\chi] = -\frac{R}{D} Q_1, \tag{3.106}$$

where the differential operator $L_s[\ ]$ is given by

$$L_s[\chi] = \frac{1}{R}\left(\chi'' + \frac{\chi'}{\varphi} - \frac{\chi}{\varphi^2}\right). \tag{3.107}$$

It will be seen that L_s can be obtained from the differential operator L of the exact theory of spherical shells (see (2.51)) if $\cot \varphi$ is replaced by $1/\varphi$ corresponding to the transition to a shallow shell.

By using the first two equations of equilibrium, we find the following expressions (which correspond to (2.22)):

$$N_{11} = -(1/\varphi)Q_1, \qquad N_{22} = -Q_1'. \tag{3.108}$$

In the compatibility condition (3.103) we now express ε_{11} and ε_{22} in terms of N_{11} and N_{22}, and the latter quantities are then expressed in terms of Q_1 by means of (3.108). In this way we find the equation

$$L_s[Q_1] = \frac{Eh}{R}\chi, \tag{3.109}$$

which corresponds to $(2.25)_2$. Eliminating χ from (3.106) and (3.109), we finally obtain

$$L_s^2[Q_1] + \frac{Eh}{D} Q_1 = 0,$$

or

$$R^2 L_s^2[Q_1] + 4\kappa^4 Q_1 = 0, \tag{3.110}$$

where

$$\kappa^4 = \frac{1}{4}\frac{Eh}{D}R^2 = 3(1-\nu^2)\frac{R^2}{h^2}. \tag{3.111}$$

It was shown in section 2.2.5 that the complete solution of the fourth-order differential equation (3.110) is determined by a linear combination of the real and imaginary parts of two linearly independent solutions of the second-order equation

$$RL_s[Q_1] - 2i\kappa^2 Q_1 = 0, \tag{3.112}$$

or, written in full,

$$Q_1'' + \frac{1}{\varphi}Q_1' - \frac{1}{\varphi^2}Q_1 - 2i\kappa^2 Q_1 = 0. \tag{3.113}$$

Let us assume that we have determined the function Q_1 by solving (3.113). The remaining statical and geometrical quantities can then be found from the equations

$$N_{11} = -Q_1/\varphi, \qquad N_{22} = -Q_1',$$

$$\chi = \frac{R}{Eh} L_s[Q_1], \qquad \mathscr{W} = \frac{R}{Eh}(\varphi Q_1' - \nu Q_1),$$

$$M_{11} = -\frac{D}{R}\left(\chi' + \nu\frac{\chi}{\varphi}\right), \qquad M_{22} = -\frac{D}{R}\left(\frac{\chi}{\varphi} + \nu\chi'\right), \tag{3.114}$$

$$\mathscr{H} = Q_1/\varphi,$$

which correspond to (2.22), $(2.25)_2$, (2.30), (2.11), (2.13), and (2.28).

If we introduce the new independent variable

$$x = (\sqrt{2})\kappa\varphi, \tag{3.115}$$

equation (3.113) assumes the form

$$x^2\frac{d^2Q_1}{dx^2} + x\frac{dQ_1}{dx} - (ix^2 + 1)Q_1 = 0. \tag{3.116}$$

This equation has the same form as (2.75). As explained in section 2.4, the solution of (3.116) is therefore given by a linear combination of the Kelvin functions of order one, $\mathrm{be}_1 x$ and $\mathrm{ke}_1 x$, while the complete solution of the fourth-order differential equation (3.110) can be represented in the form (see (2.78))

$$Q_1 = \mathrm{Re}(A_1\,\mathrm{ke}_1 x + A_2\,\mathrm{be}_1 x). \tag{3.117}$$

The Kelvin functions of order one and their derivatives can be expressed in terms of the functions of order zero and their first derivatives by means of the formulae

$$f_1 = (\sqrt{i})f_0', \qquad f_1' = (\sqrt{i})\left(if_0 - \frac{1}{x}f_0'\right), \tag{3.118}$$

where f_0 and f_1 denote either the functions $\mathrm{be}\,x$, $\mathrm{be}_1 x$, or the functions $\mathrm{ke}\,x$, $\mathrm{ke}_1 x$, and primes in connection with the Kelvin functions indicate differentiation with respect to the argument x. If we introduce (3.117) into (3.114) and use (3.118), we can express the required statical and geometrical quantities in terms of Kelvin functions of order zero in the form (2.80). Finally, the complete solution for the shallow spherical shell is obtained by adding a particular integral to the above solution of the homogeneous equations.

In this way, we find by means of calculations, the details of which will be omitted, that the complete solution can be written in the following manner:

$$\begin{bmatrix} \mathscr{H} \\ M_{11} \\ \mathscr{W} \\ \chi \\ N_{22} \\ M_{22} \\ Q_1 \end{bmatrix} = \begin{bmatrix} \mathscr{H}^{\mathrm{p}} \\ M_{11}^{\mathrm{p}} \\ \mathscr{W}^{\mathrm{p}} \\ \chi^{\mathrm{p}} \\ N_{22}^{\mathrm{p}} \\ M_{22}^{\mathrm{p}} \\ Q_1^{\mathrm{p}} \end{bmatrix} + \begin{bmatrix} (\surd 2)\kappa & & & & & & \\ & R/(\kappa\surd 2) & & & & & \\ & & R/(Eh) & & & & \\ & & & 2\kappa^2/(Eh) & & & \\ & & & & (\surd 2)\kappa & & \\ & & & & & R/(\kappa\surd 2) & \\ & & & & & & 1 \end{bmatrix} \begin{bmatrix} 0 & 0 & 1/x & 0 \\ 1 & 0 & 0 & (1-\nu)/x \\ 0 & x & -(1+\nu) & 0 \\ 0 & 0 & 0 & 1 \\ 0 & -1 & 1/x & 0 \\ \nu & 0 & 0 & -(1-\nu)/x \\ 0 & 0 & 1 & 0 \end{bmatrix} \mathbf{F}(x)\mathbf{a} \tag{3.119}$$

$$\mathbf{F}(x)\mathbf{a} = \begin{bmatrix} \ker x & \mathrm{kei}\, x & \mathrm{ber}\, x & \mathrm{bei}\, x \\ -\mathrm{kei}\, x & \ker x & -\mathrm{bei}\, x & \mathrm{ber}\, x \\ \ker' x & \mathrm{kei}'\, x & \mathrm{ber}'\, x & \mathrm{bei}'\, x \\ -\mathrm{kei}'\, x & \ker' x & -\mathrm{bei}'\, x & \mathrm{ber}'\, x \end{bmatrix} \begin{bmatrix} a_1 \\ b_1 \\ a_2 \\ b_2 \end{bmatrix}, \tag{3.119a}$$

where

$$x = (\surd 2)\kappa\varphi, \tag{3.119b}$$

and we have, according to (3.114),

$$N_{11} = -\mathscr{H}. \tag{3.119c}$$

The first column vector on the right-hand side of (3.119) denotes the particular integral. The signs ⌈ ⌋ denote a diagonal matrix. The terms containing a_1 and b_1 are damped out with increasing φ, while the terms containing a_2 and b_2 are damped out with decreasing φ (cf. the remarks on the Kelvin functions in connection with Fig. 25, section 2.4). Equation (3.119) can be written in matrix form as follows:

$$\mathbf{f} = \mathbf{f}^{\mathrm{p}}(\varphi) + \mathbf{KA}(x)\mathbf{F}(x)\mathbf{a}, \tag{3.120}$$

where the meaning of the individual matrices can be established by a comparison with (3.119).

Particular integral. It will be assumed that the conditions of section 2.2.3 are satisfied so that the *membrane solution* can be used as an approximate particular integral. Using the first two equations of equilibrium (3.104) with $Q_1 = 0$, together with (3.103) and the formula $\mathscr{W} = -R\varphi\varepsilon_{22}$, we obtain

$$\begin{aligned} N_{11}^{\mathrm{m}}(\varphi) &= -\frac{R}{\varphi^2}\int_{\varphi_0}^{\varphi} (p_3 + \varphi p_1)\varphi \,\mathrm{d}\varphi + \left(\frac{\varphi_0}{\varphi}\right)^2 N_{11}^{\mathrm{m}}(\varphi_0), \\ N_{22}^{\mathrm{m}} &= -(N_{11}^{\mathrm{m}} + Rp_3), \\ \mathscr{W}^{\mathrm{m}} &= -\frac{R\varphi}{Eh}(N_{22}^{\mathrm{m}} - \nu N_{11}^{\mathrm{m}}), \\ \chi^{\mathrm{m}} &= -\frac{1}{Eh}\left((N_{22}^{\mathrm{m}} - \nu N_{11}^{\mathrm{m}})' + \frac{1}{\varphi}(1+\nu)(N_{22}^{\mathrm{m}} - N_{11}^{\mathrm{m}})\right), \\ \mathscr{H}^{\mathrm{m}} &= -N_{11}^{\mathrm{m}}, \qquad M_{11}^{\mathrm{m}} = M_{22}^{\mathrm{m}} = Q_1^{\mathrm{m}} = 0. \end{aligned} \tag{3.121}$$

In these equations, φ_0 is the φ-value belonging to the upper edge. The expressions (3.121) should be inserted in the first column vector on the right-hand side of (3.119).

In the case of a spherical dome with a single boundary curve (parallel circle), the value $\varphi_0 = 0$ should be used in $(3.121)_1$, and the last term on the right-hand side of this equation vanishes.

Geckeler's method. In certain cases this simple approximate method can be used for the analysis of shallow spherical shells. In the derivation of Geckeler's method in section 2.6 it was assumed that the angle φ is not a very small quantity. This means that the method cannot be used for a small neighbourhood of $\varphi = 0$ (the apex of the spherical shell). However, in the case of a shallow spherical dome bounded by a single parallel circle (see Fig. 38), it is sometimes possible to use Geckeler's method for the calculation of the edge effect originating from the lower edge (the parallel circle). In this case Geckeler's method may be expected to yield acceptable results when the shell is so thin that the condition

$$1/(\varphi_e \kappa) \ll 1 \qquad (3.122)$$

is satisfied, where φ_e is the φ value belonging to the edge. When using Geckeler's method for a shallow spherical dome, the values $a_1 = b_1 = 0$ should be inserted in the formulae (2.118) in section 2.6 (since there is no upper edge), and the trigonometric functions of φ in (2.115) and (2.118) should be replaced by the approximate expressions (3.100). Moreover, the membrane solution for a shallow spherical shell should be used in the first column vector on the right-hand side of (2.118a).

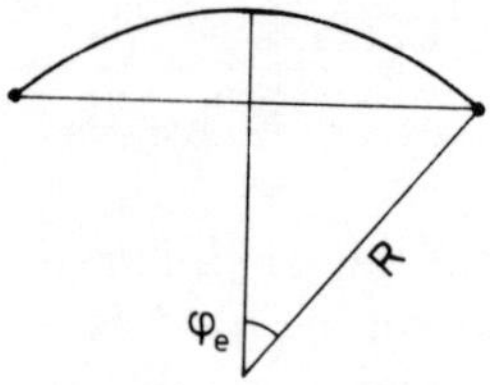

Fig. 38. Shallow spherical dome

3.6 Numerical example, shallow shell over a rectangular plan

The middle surface of the shell is an elliptic paraboloid with the equation (3.28). The relevant geometrical quantities have the following values (see also Fig. 39):

$$R_x = 48 \text{ m}, \qquad R_y = 30 \text{ m},$$
$$b = 24 \text{ m}, \qquad l = 32 \text{ m},$$
$$\text{thickness of shell, } h = 0.08 \text{ m}.$$

It is assumed that the shell is simply supported along the edges $x = \pm l/2$,

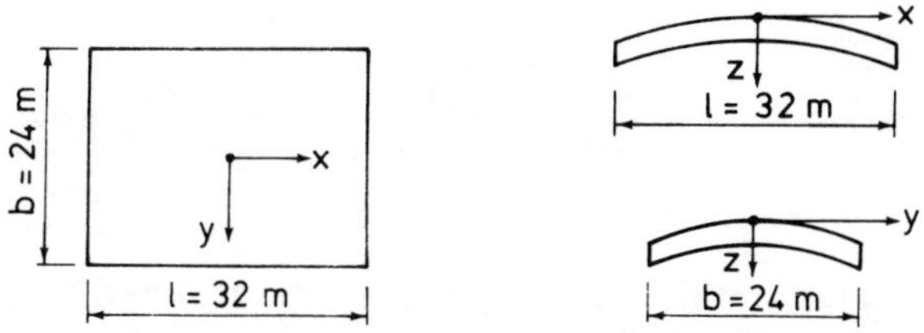

Fig. 39. Shallow shell over a rectangular plan

i.e. the corresponding edge conditions are given by (cf. (3.39))

$$\left.\begin{array}{ll} N_x = 0, & M_x = 0, \\ v_2 = 0, & w = 0, \end{array}\right\} \quad \text{for} \quad x = \pm l/2. \tag{3.123}$$

So far as the edges $y = \pm b/2$ are concerned, it is assumed that an edge beam supported by fairly closely spaced vertical columns is provided at each of these edges. The corresponding edge conditions are given approximately by equations (3.92) (for the edge $y = -b/2$). We assume that the two edge beams have the same cross-sectional area A and the same value of Young's modulus E_b.

The following three loading cases will be considered.

(a) *Loading case 1:* uniform pressure load (since the shell is shallow, the corresponding results may be regarded as an approximate solution for a uniformly distributed vertical load).

(b) *Loading case 2:* prestressing of edge beams.

(c) *Loading case 3:* the combined effect of loading cases 1 and 2 (corresponding approximately to a uniformly distributed vertical load plus prestressing of the edge beams).

We shall use the method of analysis developed in section 3.3. The uniformly distributed pressure load p is expanded in a Fourier half-range series in the X-direction. Thus

$$p = p_0 = \frac{4}{\pi}\, p_0\left[\cos\left(\pi\frac{x}{l}\right) - \frac{1}{3}\cos\left(3\pi\frac{x}{l}\right) + \frac{1}{5}\cos\left(5\pi\frac{x}{l}\right) - \ldots\right], \tag{3.124}$$

where p_0 is a constant. As explained in section 3.3, the solution is determined as the sum of the solutions for each of the terms in the Fourier series. In the present example we shall restrict ourselves to a derivation of the solution for the first term of the series. We then have (cf. (3.37b)

$$\alpha_1 = \frac{\pi}{l} = \frac{\pi}{32}\,\text{m}^{-1}.$$

Solution of auxiliary equation. Assuming that Poisson's ratio is zero, we find by substituting the numerical values in the formulae (3.52a), (3.56),

(3.60), and (3.61):

$$k = 1.6,$$
$$\gamma = 46.7983, \qquad \beta = 56.1580, \qquad (3.125a)$$
$$s = 0.599\,951, \qquad t = 46.8021.$$

Inserting these values into (3.65), we find for the real and imaginary parts of the roots of the auxiliary equation:

$$\begin{aligned} j_1 &= 1.264\,893, & k_1 &= 0.001\,5021, \\ j_2 &= 6.855\,70, & k_2 &= 6.826\,46. \end{aligned} \qquad (3.125b)$$

Determination of arbitrary constants. The edge conditions at the edge $y = -b/2$ are expressed in terms of the dimensionless functions (3.45) by the following equations (see (3.93)):

$$\left.\begin{aligned} \tilde{w} = 0, \quad \tilde{N}_y = 0, \quad \tilde{M}_y = 0, \\ \tilde{N}_{xy} - A\frac{\alpha_1}{h}\tilde{v}_1 = 0, \end{aligned}\right\} \quad \text{for} \quad \eta = -\tfrac{1}{2}\alpha_n b, \qquad (3.126)$$

where it is assumed that Young's moduli for the shell and the edge beam have the same value, $E = E_b$.

If we introduce the vector

$$\mathbf{X}^{\mathrm{T}} = [\tilde{N}_y \quad \tilde{M}_y \quad \tilde{w} \quad \tilde{N}_{xy} \quad \tilde{v}_1], \qquad (3.127)$$

it follows from (3.88) that $\mathbf{X}$ can be written in the form

$$\mathbf{X} = \mathbf{X}_0 + \mathbf{CF}(\zeta)\mathbf{a} + \mathbf{J}_1\mathbf{CF}(\theta)\mathbf{b}, \qquad (3.128)$$

where

$$\mathbf{C} = \begin{bmatrix} -1 & 0 & -1 & 0 \\ (t-\gamma) & (1+s) & -(t+\gamma) & -(1-s) \\ 0 & -1 & 0 & 1 \\ -j_1 & k_1 & -j_2 & k_2 \\ (1+s) & -(t+\gamma) & (1-s) & -(t-\gamma) \end{bmatrix}, \qquad \mathbf{J}_1 = \begin{bmatrix} 1 & & & & \\ & 1 & & & \\ & & 1 & & \\ & & & -1 & \\ & & & & 1 \end{bmatrix} \qquad (3.128a)$$

$\mathbf{J}_1$ is a diagonal matrix, and $\mathbf{X}_0$ is the contribution due to the particular integral and the prestressing of the edge beams. It will appear from the following that $\mathbf{X}_0$ is a constant vector for each of the above loading cases. The matrix $\mathbf{C}$ is obtained from Table 3.2 by selecting the appropriate rows. Formulae for $\mathbf{X}_0$ will be given in the following.

As the shell, the supports, and the loading are all symmetrical with respect to $y = 0$, we conclude that the solution will also be symmetrical with respect to $y = 0$. It follows that the arbitrary constants satisfy the condition

$$\mathbf{a} = \mathbf{b}, \qquad (3.129)$$

since it can be seen from (3.128) that the appropriate conditions of symmetry will be satisfied by each of the components of $\mathbf{X}$ when equation (3.129) holds, and when we use a particular integral which satisfies the symmetry conditions.

Using (3.128) we now obtain for the left-hand edge $y = -b/2$ (i.e. $\zeta = 0$, $\theta = \alpha_1 b$):

$$\mathbf{X}_{\zeta=0} = \mathbf{X}_0 + [\mathbf{C} + \mathbf{J}_1\mathbf{C}\mathbf{F}(\alpha_1 b)]\mathbf{a}, \tag{3.130}$$

since $\mathbf{F}(0) = \mathbf{I}$. The edge conditions (3.126) can be written in matrix form as

$$\mathbf{Q}\mathbf{X}_{\zeta=0} = \mathbf{0}, \tag{3.131}$$

$$\mathbf{Q} = \begin{bmatrix} 1 & 0 & 0 & 0 & 0 \\ 0 & 1 & 0 & 0 & 0 \\ 0 & 0 & 1 & 0 & 0 \\ 0 & 0 & 0 & 1 & -A(\alpha_1/h) \end{bmatrix}. \tag{3.131a}$$

Inserting (3.130) into (3.131), we obtain

$$\mathbf{Q}\{\mathbf{X}_0 + [\mathbf{C} + \mathbf{J}_1\mathbf{C}\mathbf{F}(\alpha_1 b)]\mathbf{a}\} = \mathbf{0}. \tag{3.132}$$

This is a system of four linear equations with four unknowns for the determination of the arbitrary constants.

Loading case 1. In this case the particular integral is given by Table 3.3 in section 3.3.3, and the constant $\tilde{p}$ has the value

$$\tilde{p} = \frac{p_1}{p_0} = \frac{4}{\pi}$$

(see (3.85c) and (3.124)). By selecting the appropriate components from Table 3.3, we get

$$\mathbf{X}_0^{\mathrm{T}} = \begin{bmatrix} -\tilde{\psi}^{\mathrm{P}} & 0 & \dfrac{\tilde{\psi}^{\mathrm{P}}}{2\gamma k} & 0 & \dfrac{\tilde{\psi}^{\mathrm{P}}}{k} \end{bmatrix}, \tag{3.133}$$

where

$$\tilde{\psi}^{\mathrm{P}} = \frac{4\gamma^2 k^2}{1 + 4\gamma^2 k^2}\tilde{p}. \tag{3.133a}$$

Let the cross-sectional area of the edge beam be $A = 0.100\ \mathrm{m}^2$. Then we have

$$A\frac{\alpha_1}{h} = 0.122\,719.$$

By inserting the numerical values in (3.132), we find the following system of linear equations:

$$\begin{bmatrix} -1.050\,776 & -0.000\,180 & -1 & 0 \\ 0.003\,705 & 1.681\,190 & -93.6004 & -0.400\,046 \\ 0.000\,180 & -1.050\,776 & 0 & 1 \\ -1.409\,044 & 12.071\,35 & -6.904\,79 & 6.826\,93 \end{bmatrix} \begin{bmatrix} a_1 \\ b_1 \\ a_2 \\ b_2 \end{bmatrix} = \begin{bmatrix} 1.273\,183 \\ 0 \\ -0.008\,502 \\ 0.097\,652 \end{bmatrix}. \tag{3.134}$$

These equations have the solution

$$\begin{bmatrix} a_1 \\ b_1 \\ a_2 \\ b_2 \end{bmatrix} = \begin{bmatrix} -1.210\,595 \\ -8.101\,86 \times 10^{-2} \\ -1.103\,863 \times 10^{-3} \\ -9.341\,65 \times 10^{-2} \end{bmatrix}. \tag{3.134a}$$

We shall now determine the contact forces and couples N_x, N_y, N_{xy}, and M_y. The corresponding dimensionless functions can be computed from (3.88). Using the values (3.134a) of the arbitrary constants, we obtain the results shown in Table 3.4. The quantity υ in this table is proportional to the distance from the left-hand edge, i.e.

$$\upsilon = (y + b/2)/b.$$

Table 3.4 Loading case 1

υ	$\tilde{N}_x$	$\tilde{N}_y$	$\tilde{N}_{xy}$	$\tilde{M}_y$
0	6.708	0	0.823	0
0.05	0.877	−0.128	1.242	2.729
0.10	−1.663	−0.273	1.165	1.628
0.15	−1.979	−0.397	0.938	0.412
0.30	−1.024	−0.628	0.431	−0.138
0.50	−0.874	−0.727	0	−0.054

Because of the symmetry, results are given for half the shell only. The variation of the contact forces and couples in the X-direction is determined by (3.45) and Table 3.1. Graphs of the contact forces and couples are shown in Fig. 40a.

Loading case 2. It is assumed that the prestressing cables run the whole length of the beam and are anchored at the ends, and that both edge beams are identically prestressed. If we imagine that the beam is separated from the shell by a section along the edge, the prestress would produce a constant direct force $-N_0$ (compression) throughout the beam. This direct force is expanded in a Fourier half-range series in the X-direction

$$N_{\text{pr}} = -N_0 = -\frac{4}{\pi} N_0 \left[\cos\left(\pi \frac{x}{l}\right) - \frac{1}{3} \cos\left(3\pi \frac{x}{l}\right) + \dots \right],$$

where the subscript pr denotes the contribution from the prestress. In the following we confine ourselves to the solution for the first term of the series. We therefore assume that the prestressing force is given by

$$N_{\text{pr}} = -\frac{4}{\pi} N_0 \cos\left(\pi \frac{x}{l}\right). \tag{3.135}$$

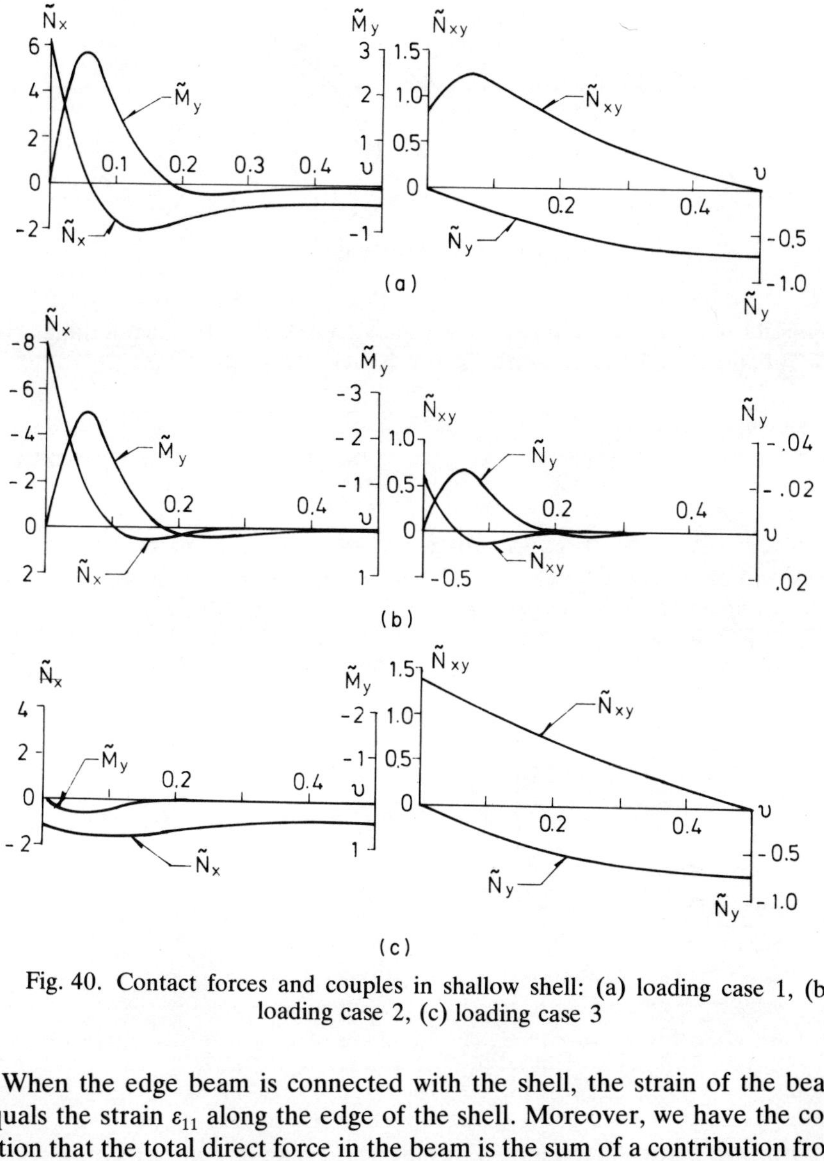

Fig. 40. Contact forces and couples in shallow shell: (a) loading case 1, (b) loading case 2, (c) loading case 3

When the edge beam is connected with the shell, the strain of the beam equals the strain ε_{11} along the edge of the shell. Moreover, we have the condition that the total direct force in the beam is the sum of a contribution from the prestressing and a contribution from the shear forces N_{xy} in the shell. For the left-hand edge, we therefore have (see Fig. 41)

$$\frac{dN}{dx} = \frac{dN_{pr}}{dx} - N_{xy}, \qquad N = E_b A \frac{\partial v_1}{\partial x}.$$

Eliminating N, we obtain the condition

$$E_b A \frac{\partial^2 v_1}{\partial x^2} = \frac{dN_{pr}}{dx} - N_{xy}. \tag{3.136}$$

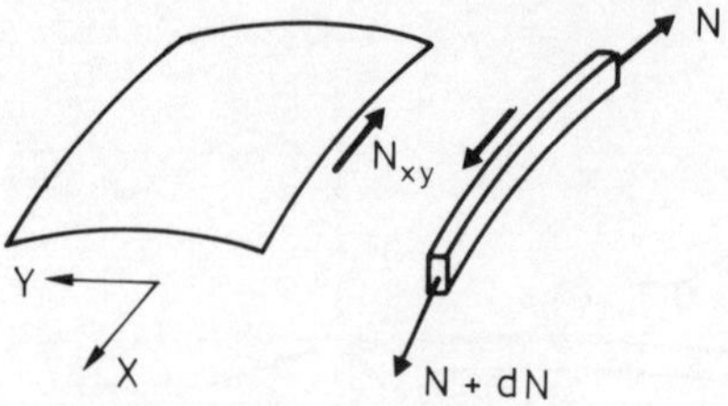

Fig. 41. Forces on edge beam

In addition, we have the three conditions (3.92a). By introducing dimensionless functions, the edge conditions can be written in the form

$$\tilde{w} = 0, \qquad \tilde{N}_y = 0, \qquad \tilde{M}_y = 0,$$
$$\tilde{N}_{xy} - A\frac{\alpha_1}{h}\tilde{v}_1 - \frac{4}{\pi}\frac{\alpha_1}{p_0 R_y}N_0 = 0. \tag{3.137}$$

Since $p = 0$ in the present loading case, the particular integral vanishes, so there will be no contribution to $\mathbf{X}_0$ from this source. However, there will be a contribution to $\mathbf{X}_0$ from the prestressing. The edge conditions (3.137) can be written in the form (see (3.131))

$$\mathbf{Q}\begin{bmatrix} \tilde{N}_y \\ \tilde{M}_y \\ \tilde{w} \\ \tilde{N}_{xy} - \dfrac{4}{\pi}\dfrac{\alpha_1 N_0}{p_0 R_y} \\ \tilde{v}_1 \end{bmatrix} = \mathbf{0},$$

or

$$\mathbf{Q}(\mathbf{X} + \mathbf{X}_0) = \mathbf{0} \qquad \text{for} \qquad \eta = -\tfrac{1}{2}\alpha_1 b, \tag{3.138}$$

where

$$\mathbf{X}_0^{\mathrm{T}} = \begin{bmatrix} 0 & 0 & 0 & -\dfrac{4}{\pi}\dfrac{\alpha_1 N_0}{p_0 R_y} & 0 \end{bmatrix}. \tag{3.138a}$$

Since the particular integral vanishes, the vector $\mathbf{X}$ consists of the contribution from the solution of the homogeneous equations only and is therefore given by the last two terms on the right-hand side of (3.128). By inserting this expression for $\mathbf{X}$ in (3.138), we again obtain an equation of the form (3.132), but $\mathbf{X}_0$ is now given by (3.138a).

By assuming the value

$$N_0 = 0.38 p_0 R_y l, \tag{3.139}$$

for the prestressing force, we find a system of linear equations for the determination of the arbitrary constants, whose matrix of coefficients is identical with the matrix on the left-hand side of (3.134), while the right-hand side, by means of (3.138), is found to be

$$\begin{bmatrix} 0 \\ 0 \\ 0 \\ 1.52 \end{bmatrix}$$

By solving these equations, we find the following values of the arbitrary constants:

$$\begin{bmatrix} a_1 \\ b_1 \\ a_2 \\ b_2 \end{bmatrix} = \begin{bmatrix} -1.029\,991 \times 10^{-3} \\ 7.928\,96 \times 10^{-2} \\ 1.068\,018 \times 10^{-3} \\ 8.331\,58 \times 10^{-2} \end{bmatrix} \tag{3.140}$$

The contact forces and couples can now be computed. The resulting values of the dimensionless functions are given in Table 3.5, and corresponding graphs are shown in Fig. 40b.

Loading case 3. For this loading case (uniformly distributed load plus prestressing) the results are obtained by superposition of the results from the two preceding loading cases. The computed values of the dimensionless functions are given in Table 3.6, and corresponding graphs are shown in Fig. 40c.

Table 3.5 Loading case 2

υ	$\tilde{N}_x$	$\tilde{N}_y$	$\tilde{N}_{xy}$	$\tilde{M}_y$
0	−7.800	0	0.563	0
0.05	−2.371	−0.026	−0.011	−2.429
0.10	0.083	−0.016	−0.118	−1.445
0.15	0.522	−0.004	−0.071	−0.360
0.30	−0.009	0.001	0.005	0.129
0.50	0	0	0	0.052

Table 3.6 Loading case 3

υ	$\tilde{N}_x$	$\tilde{N}_y$	$\tilde{N}_{xy}$	$\tilde{M}_y$
0	−1.092	0	1.386	0
0.05	−1.494	−0.154	1.231	0.300
0.10	−1.580	−0.289	1.047	0.183
0.15	−1.457	−0.401	0.867	0.052
0.30	−1.033	−0.627	0.436	−0.009
0.50	−0.874	−0.727	0	−0.002

By comparing the results of loading cases 1 and 3 it will be seen that the large tensile forces N_x at the edge in the shell without prestress are replaced by moderate compressive forces in the prestressed shell, and that the bending moments M_y are reduced considerably.

It should be noted that in practical applications of the above analytical solution, one would generally include more than one term of the Fourier series.

3.7 Numerical example, shallow spherical shell

We now consider a spherical shell of constant thickness (see Fig. 42). The shell is bounded by a single edge curve (the parallel circle $\varphi = \varphi_1$), and a ring beam is provided along this edge. The relevant geometrical parameters have the following values:

$$R = 18 \text{ m}, \qquad h = \frac{\sqrt{3}}{25} = 0.069\,282 \text{ m}, \qquad \varphi_1 = \frac{1}{3}.$$

The loading consists of a uniform pressure load, and it is assumed that Poisson's ratio is zero ($\nu = 0$).

We shall first determine the contact forces and couples in the shell by solving the governing equations of the theory of shallow, spherical shells, and we shall then calculate an approximate solution for the shallow shell by applying Geckeler's method.

The loading is given by

$$p_1 = 0, \qquad p_3 = p_0, \tag{3.141}$$

where p_0 is the constant pressure. We begin by determining the *membrane solution*. By substituting the expressions (3.141) for the load in (3.121), we obtain (since the last term in the formula for N_{11}^{m} vanishes; cf. the remarks following equations (3.121):

$$\begin{aligned} &N_{11}^{\mathrm{m}} = N_{22}^{\mathrm{m}} = -\tfrac{1}{2}p_0 R, \\ &\mathscr{H}^{\mathrm{m}} = \tfrac{1}{2}p_0 R, \qquad M_{11}^{\mathrm{m}} = 0, \\ &\mathscr{W}^{\mathrm{m}} = \frac{1}{2}\frac{p_0 R^2}{Eh}\varphi. \end{aligned} \tag{3.142}$$

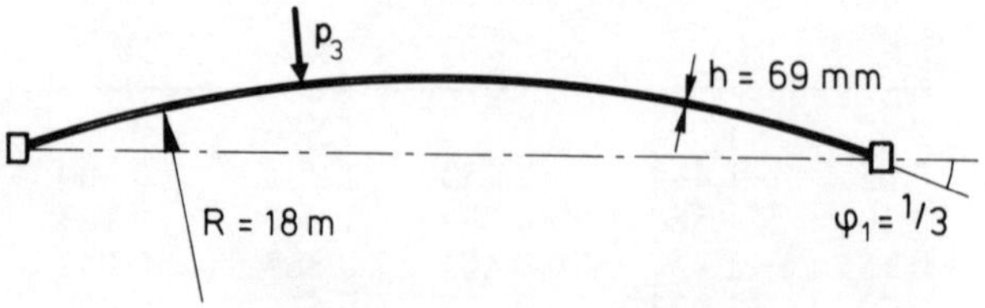

Fig. 42. Shallow spherical shell

It will be convenient to introduce *dimensionless quantities* defined in the following manner:

$$\tilde{N}_{11} = N_{11}/(p_0R), \qquad \tilde{N}_{22} = N_{22}/(p_0R),$$
$$\tilde{\mathscr{H}} = \mathscr{H}/(p_0R), \qquad \tilde{M}_{11} = M_{11}/(p_0R^2), \tag{3.143}$$
$$\tilde{\mathscr{W}} = \frac{Eh}{p_0R^2}\mathscr{W}.$$

The membrane solution can then be written in the dimensionless form

$$\tilde{N}_{11}^{\mathrm{m}} = \tilde{N}_{22}^{\mathrm{m}} = -1/2, \qquad \tilde{M}_{11}^{\mathrm{m}} = 0,$$
$$\tilde{\mathscr{H}}^{\mathrm{m}} = 1/2, \qquad \tilde{\mathscr{W}}^{\mathrm{m}} = \varphi/2. \tag{3.144}$$

(a) *Solution of governing equations for shallow spherical shell*. We first apply the equations of section 3.5. We have

$$R/h = 150\sqrt{3}$$

and inserting this value in (3.111) and (3.115), we obtain

$$\kappa = \sqrt{[(\sqrt{3})R/h]} = \sqrt{450} = 21.2132, \tag{3.145}$$

$$x = (\sqrt{2})\kappa\varphi = 30\varphi. \tag{3.146}$$

Edge conditions. In order to formulate the edge conditions, we shall use the results from section 2.2.4 concerning edge beams. The conditions at the edge of the spherical shell correspond to the case in which only one segment of the shell occurs in Fig. 23 (namely, the upper segment 1). As explained in section 2.2.4, we then have the following two edge conditions corresponding to equations $(2.44)_{3,4}$ (since $\mathscr{H}_2 = \bar{P}_{\mathrm{h}} = 0$ in the present case):

$$\mathscr{H} = -\frac{E_{\mathrm{b}}A}{r_0^2}\mathscr{W}, \qquad M_{11} = 0. \tag{3.147}$$

We shall impose the condition that the contact forces and couples shall remain finite throughout the shell (and, in particular, at the apex). Let us now consider the expression (3.119) for the complete solution. We note that the functions $\ker x$, $\mathrm{kei}\, x$, $\ker' x$ and $\mathrm{kei}' x$ become infinite at the apex (i.e. for $x = 0$), and that these functions appear only in the coefficients of the arbitrary constants a_1 and b_1. It is easily verified that the said coefficients likewise become infinite at the apex. In order that the contact forces and couples may remain finite at the apex, we must therefore put

$$a_1 = b_1 = 0. \tag{3.148}$$

By introducing the expressions for $\mathscr{H}$, $\mathscr{W}$, and M_{11} from (3.119) in the conditions (3.147), we now obtain

$$\mathscr{H}^{\mathrm{p}} + (\sqrt{2})\kappa \frac{1}{x}[(\mathrm{ber}' \, x)a_2 + (\mathrm{bei}' \, x)b_2]$$

$$= -\frac{E_b A}{r_0^2}\left(\mathscr{W}^{\mathrm{p}} + \frac{R}{Eh}[-(x \,\mathrm{bei}\, x + \mathrm{ber}' \, x)a_2 + (x \,\mathrm{ber}\, x - \mathrm{bei}' \, x)b_2]\right),$$

$$M_{11}^{\mathrm{p}} + \frac{R}{(\sqrt{2})\kappa}\left[\left(\mathrm{ber}\, x - \frac{1}{x}\mathrm{bei}' \, x\right)a_2 + \left(\mathrm{bei}\, x + \frac{1}{x}\mathrm{ber}' \, x\right)b_2\right] = 0. \quad (3.149)$$

If we use the membrane solution as an approximate, particular integral, and introduce the dimensionless quantities (3.143) and the dimensionless arbitrary constants

$$[\tilde{a}_2 \quad \tilde{b}_2] = \frac{1}{p_0 R}[a_2 \quad b_2], \quad (3.150)$$

the conditions (3.149) will assume the form

$$\frac{1}{2} + \frac{(\sqrt{2})\kappa}{x}[(\mathrm{ber}' \, x)\tilde{a}_2 + (\mathrm{bei}' \, x)\tilde{b}_2]$$

$$= -\frac{AR}{r_0^2 h}[\varphi/2 - (x \,\mathrm{bei}\, x + \mathrm{ber}' \, x)\tilde{a}_2 + (x \,\mathrm{ber}\, x - \mathrm{bei}' \, x)\tilde{b}_2], \quad (3.151)$$

$$\left(\mathrm{ber}\, x - \frac{1}{x}\mathrm{bei}' \, x\right)\tilde{a}_2 + \left(\mathrm{bei}\, x + \frac{1}{x}\mathrm{ber}' \, x\right)\tilde{b}_2 = 0,$$

where $\varphi = \varphi_1 = 1/3$ and $x = 10$ at the edge, and it is assumed that the values of Young's modulus for the shell and the edge beam are the same.

Let the cross-sectional area of the edge beam be

$$A = 0.0180 \text{ m}^2,$$

and note that

$$r_0 = R \sin \varphi_1 \simeq R\varphi_1 = 6.00 \text{ m}.$$

Inserting these numerical values in (3.151), we find the linear equations

$$\begin{aligned} 73.72\tilde{a}_2 + 568.6\tilde{b}_2 &= -0.5217, \\ 125.3\tilde{a}_2 + 61.49\tilde{b}_2 &= 0, \end{aligned} \quad (3.152)$$

the solution of which is given by

$$\tilde{a}_2 = 0.480\,91 \times 10^{-3}, \qquad \tilde{b}_2 = -0.979\,73 \times 10^{-3}. \quad (3.153)$$

The values of the Kelvin functions have been taken from Hütte,[41] Table 24, p. 64.

Contact forces and couples. The contact forces and couples can now be computed by means of the representation (3.119). We use the membrane solution as a particular integral, introduce dimensionless quantities, and insert the values (3.153) of the arbitrary constants. The results are given in Table 3.7 and are plotted graphically in Fig. 43, the independent variable υ being proportional to the arc length measured from the edge, i.e.

$$\upsilon = \frac{1}{\varphi_1}(\varphi_1 - \varphi). \qquad (3.154)$$

It will be seen that one may distinguish between an inner region ($0.5 \leqslant \upsilon \leqslant 1$), in which the behaviour of the shell is closely described by membrane theory, and an edge zone ($0 \leqslant \upsilon \leqslant 0.5$), in which the edge effects are significant.

Table 3.7 Contact forces and couples

	Exact solution			Geckeler's method		
υ	$\tilde{N}_{11}$	$\tilde{N}_{22}$	$\tilde{M}_{11} \times 10^3$	$\tilde{N}_{11}$	$\tilde{N}_{22}$	$\tilde{M}_{11} \times 10^3$
0	−0.1762	4.0691	0	−0.1765	4.0754	0
0.1	−0.4866	1.3034	1.6896	−0.4804	1.2151	1.6284
0.2	−0.5972	−0.2917	1.3389	−0.5818	−0.3265	1.2208
0.3	−0.5936	−0.8196	0.6127	−0.5762	−0.7869	0.5194
0.4	−0.5524	−0.8138	0.1282	−0.5401	−0.7573	0.0926
0.5	−0.5145	−0.6670	−0.0647	−0.5102	−0.6231	−0.0569
0.6	−0.4923	−0.5000	−0.0884	−0.4949	−0.5298	−0.0651
0.7	−0.4838	−0.4947	−0.0544	−0.4908	−0.4924	−0.0350

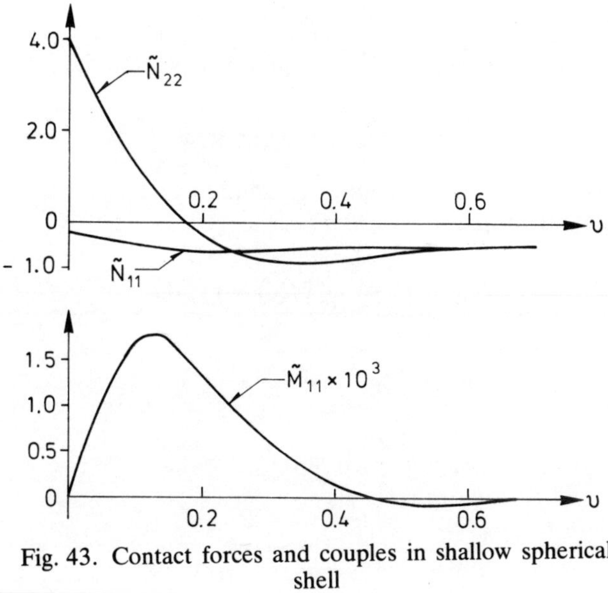

Fig. 43. Contact forces and couples in shallow spherical shell

(b) *Geckeler's method*. We shall now determine an approximate solution by applying Geckeler's method to the shallow shell (see section 3.5). Since the value

$$\frac{1}{\varphi_e\kappa} = \frac{3}{21.21} = \frac{1}{7.07}$$

is small compared with unity, we may expect that the results obtained by Geckeler's method will be acceptable (cf. the remarks preceding formula (3.122)). It was explained in section 3.5 that we must put $a_1 = b_1 = 0$ in (2.118) in the present case because of the absence of an edge effect originating from the upper edge.

For the spherical shell we have the relations

$$R_{2u} = R_{21} = R.$$

The formulae (2.113), (2.114a), and (2.116a) therefore assume the form

$$\zeta = \frac{1}{R}(s_1 - s) = \varphi_1 - \varphi = \varphi_1 \upsilon, \tag{3.155}$$

$$\kappa_u = \kappa_1 = \kappa = \sqrt{[(\sqrt{3})R/h]} = 21.2132. \tag{3.156}$$

In the equations (2.115) and (2.118), we now insert the approximations (3.100) for the trigonometric functions of φ, and we also introduce the dimensionless quantities (3.143) and (3.150). Since $f(0) = 1$ and $g(0) = 0$, the edge conditions assume the form

$$\begin{gathered}\frac{1}{2} + \frac{1}{\varphi_1}\tilde{a}_2 = -\frac{AR}{r_0^2 h}\left(\frac{\varphi_1}{2} + \kappa\varphi_1(\tilde{a}_2 - \tilde{b}_2)\right), \\ \tilde{a}_2 + \tilde{b}_2 = 0.\end{gathered} \tag{3.157}$$

Inserting the numerical values in these equations, we find the following values of the arbitrary constants:

$$\tilde{a}_2 = -\tilde{b}_2 = -0.107\,843. \tag{3.158}$$

From (2.118) we obtain the following simple expressions for the required contact forces and couples:

$$\begin{aligned}\tilde{N}_{11} &= -\frac{1}{2} - \frac{1}{\varphi}[f(\zeta) - g(\zeta)]\tilde{a}_2, \\ \tilde{N}_{22} &= -\frac{1}{2} - 2\kappa f(\zeta)\tilde{a}_2, \\ \tilde{M}_{11} &= -\frac{1}{\kappa}g(\zeta)\tilde{a}_2,\end{aligned} \tag{3.159}$$

where

$$f(\zeta) = e^{-\kappa\zeta}\cos\kappa\zeta, \qquad g(\zeta) = e^{-\kappa\zeta}\sin\kappa\zeta.$$

The calculated values are given in Table 3.7. It will be seen that the agreement between the previous results and the approximate results according to Geckeler's method is fairly good. The greatest relative deviation between the two sets of results occurs in the values of the bending moment and amounts to 6.5% of the maximum value.

Problems

3.1 The equation of the middle surface of a shallow shell is given by

$$z = \frac{1}{2}\left(\frac{x^2}{R_x} + \frac{y^2}{R_y}\right),$$

where R_x and R_y are constants, and $R_x R_y \neq 0$. The projection of the edge curve on the XY-plane is a rectangle whose sides are given by $x = \pm l/2$ and $y = \pm b/2$ (see Fig. 32 in section 3.3.3).

In the solution of the present problem, the bending and torsional moments as well as the transverse shear forces should be neglected in the equations for the shallow shell, i.e. the *membrane theory* should be used for the shallow shell.

Establish the governing differential equation for the stress function ψ. Determine ψ and the membrane contact forces N_x, N_y, and N_{xy} for the following load:

$$p_1 = p_2 = 0, \qquad p = p_0 \cos \alpha x, \qquad \alpha = \frac{n\pi}{l},$$

where p_0 is a constant, and n is an odd, positive integer. The edge conditions are given by

$$N_x = 0 \quad \text{for} \quad x = \pm l/2,$$
$$N_y = 0 \quad \text{for} \quad y = \pm b/2.$$

3.2 Consider a shallow shell over a rectangular plan as described in problem 3.1.

(1) Using the expressions for the strain measures of shallow shell theory, derive a differential equation for w that can be used to calculate the normal component of *inextensional displacements* of the shell (i.e. displacements for which the strain measures ε_{11}, ε_{22}, and ε_{12} are zero everywhere on the middle surface).

(2) By solving the equation derived in (1), determine the normal component w of the inextensional displacement when the edge conditions are given by

$$w = 0 \quad \text{for} \quad x = \pm l/2,$$
$$w = A\cos\left(\frac{\pi x}{l}\right) \quad \text{for} \quad y = b/2,$$
$$w = B\cos\left(\frac{\pi x}{l}\right) \quad \text{for} \quad y = -b/2,$$

where A and B are constants. It is also assumed that the constants R_x and R_y in the equation of the middle surface satisfy the condition $R_x \geqslant R_y > 0$.

(3) Determine the corresponding tangential components v_1 and v_2 of the inextensional displacement, assuming that

$$v_1 = 0 \quad \text{for} \quad x = 0,$$
$$v_2 = 0 \quad \text{for} \quad x = \pm l/2.$$

3.3 The middle surface of a shallow shell of constant thickness is given by the equation

$$z = \frac{1}{2R}(x^2 + y^2),$$

where R is a positive constant. The projection of the edge curve on the XY-plane is a square whose sides are given by $x = \pm l/2$ and $y = \pm l/2$. It is assumed that the shell is elastic and that Poisson's ratio is zero.

(1) The applied surface load acting on the shell is given by

$$p_1 = p_2 = 0, \qquad p = p_0 \cos\left(\frac{\pi x}{l}\right)\cos\left(\frac{\pi y}{l}\right),$$

where p_0 is a constant. Assume a solution of the form

$$\psi = \psi_0 \cos\left(\frac{\pi x}{l}\right)\cos\left(\frac{\pi y}{l}\right), \qquad w = w_0 \cos\left(\frac{\pi x}{l}\right)\cos\left(\frac{\pi y}{l}\right),$$

where ψ_0 and w_0 are constants, and determine the contact forces N_x, N_y, N_{xy}, Q_x, and Q_y in the shell.

(2) Using the approximations $\cos(n, z) = 1$, $\sin \varphi = x/R$, where (n, z) is the angle between the surface normal at an arbitrary point of the middle surface and the Z-axis, and φ is the angle between the tangent to the curve y = constant on the middle surface and the X-axis, show that the condition of force equilibrium in the direction of the Z-axis is satisfied for the shell as a whole (the resulting equilibrium equation will contain contributions from the applied surface load and from the contact forces on the edges).

(3) Use the above equilibrium equation to determine the ratio of the part of the applied load carried by the shear forces Q_x and Q_y (i.e. by bending and torsion) to the total applied load on the shell, when

(a) $R = l, \qquad h/l = 1/20,$
(b) $1/R = 0, \qquad h/l = 1/20.$

Case (b) corresponds to a plane, square plate situated in the XY-plane.

3.4 Consider a cylindrical tube of length l and of constant thickness h. The radius of the circular cross-section of the middle surface is denoted by R. The axis of the tube is horizontal, and the angle θ (see Fig. 3A) is measured from the vertical direction.

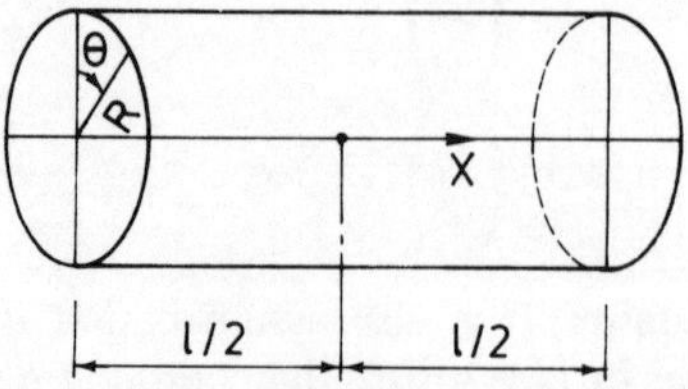

Fig. 3A

The tube is regarded as an elastic, cylindrical shell with Poisson's ratio $\nu = 0$, and the Donnell theory is used.

(1) The shell is subjected to the following applied surface load:

$$p_1 = p_2 = 0, \qquad p = p_0 \cos\left(\frac{\pi x}{l}\right) \cos\theta,$$

where p_0 is a constant. Assume a solution of the form

$$\psi = \psi_0 \cos\left(\frac{\pi x}{l}\right) \cos\theta, \qquad w = w_0 \cos\left(\frac{\pi x}{l}\right) \cos\theta,$$

where ψ_0 and w_0 are constants, and determine the corresponding contact forces N_x, N_{xy}, and Q_x in the shell.

(2) Determine the resultant of the forces acting on the tube, i.e. the applied surface load and the contact forces on the end sections (use the contact forces and not the reduced edge loads in this calculation). Show that this resultant force does not vanish, and explain why this is so. Show that the ratio between the magnitudes of the resultant force and the vertical component of the total applied load equals 1/127 when $l = \pi R$ and $R = 10(\sqrt{\tfrac{5}{6}})h$.

3.5 Consider a shallow cylindrical shell of constant thickness, for which the geometrical parameters have the values (see (3.28) and Fig. 32):

$$l = R_y = 24 \text{ m}, \qquad 1/R_x = 0,$$

$$h = \frac{3\sqrt{3}}{8\pi^2} \simeq 0.066 \text{ m}.$$

The shell is elastic, and Poisson's ratio is zero. The applied surface load is given by

$$p_1 = p_2 = 0, \qquad p = p_0 \cos\left(\frac{\pi x}{l}\right),$$

where p_0 is a constant. The shell is simply supported at the edges $x = \pm l/2$, and the following edge conditions (flexible edge beams) are valid at the edges $y = \pm b/2$:

$$N_y = 0, \qquad M_y = 0, \qquad Q_y + M_{xy,x} = 0,$$

$$EA\,\frac{\partial^2 v_1}{\partial x^2} \mp N_{xy} = 0,$$

where the upper sign applies at the edge $y = b/2$, and the lower sign at the edge $y = -b/2$. E is Young's modulus (the same value applies to both the edge beam and the shell), A is the cross-sectional area of the edge beam, and we have

$$\frac{\pi A}{lh} = 0.10.$$

(1) Compute the roots of the auxiliary equation (see section 3.3.1).

(2) Formulate the equations that determine the arbitrary constants associated with the left-hand edge ($y = -b/2$). Assume for this purpose that the influence of the edge effect from the opposite edge can be neglected, and use the approximation

$$\tilde{\psi}^{\mathrm{p}} = \tilde{p}$$

for the constant (3.87a) appearing in the particular integral.

3.6 For a thin, shallow shell over a rectangular plan, it can be shown that the dimensionless constants appearing in the solution of the homogeneous differential equation for the stress function (see section 3.3.1) are given approximately by

$$s \simeq k - 1, \qquad t \simeq \gamma,$$
$$j_1 \simeq \sqrt{k}, \qquad k_1 \simeq 0,$$
$$j_2 \simeq \sqrt{\gamma}, \qquad k_2 \simeq \sqrt{\gamma}.$$

It is assumed here that the shell is elastic and of constant thickness, that Poisson's ratio is zero, and that the shell is simply supported at the edges $x = \pm l/2$.

A shallow shell of this type is subjected to the following applied surface load:

$$p_1 = p_2 = 0, \qquad p = p_0 \cos\left(\frac{\pi x}{l}\right),$$

where p_0 is a constant. The edge conditions at the edges $y = \pm b/2$ are given by

$$N_y = 0, \quad M_y = 0, \quad w = 0, \quad v_1 = 0,$$

and the dimensionless constants k and γ have the values

$$k = 2, \qquad \gamma = 49.$$

(1) Formulate the equations that determine the arbitrary constants associated with the left-hand edge ($y = -b/2$), using the above approximate values of the dimensionless constants and neglecting the influence of the edge effect from the opposite edge. The approximate value

$$\tilde{\psi}^{\mathrm{p}} = \tilde{p}$$

should be used for the constant appearing in the particular integral.

(2) Solve these equations to find the values of the arbitrary constants associated with the left-hand edge (begin by eliminating the arbitrary constant a_1 from the two equations that express the conditions $N_y = 0$ and $v_1 = 0$).

(3) Determine the value of $Q_y + M_{xy,x}$ (expressed as a multiple of $p_0 l$) at the point $(x, y) = (0, \; -b/2)$.

3.7 The dimensions of a shallow, spherical dome of constant thickness are given as follows:

$$R = 15.00 \text{ m}, \qquad h = 0.10 \text{ m}, \qquad \varphi_e = 0.4,$$

where φ_e is the φ value belonging to the edge. The shell is assumed to be elastic, and Poisson's ratio is assumed to be zero.

The shell is subjected to a constant pressure load, i.e.

$$p_1 = p_2 = 0, \qquad p_3 = p_0 = \text{constant}.$$

Use Geckeler's method to determine the values of the arbitrary constants for this load and for the following two types of edge conditions:

(1) $\mathscr{H} = 0, \quad M_{11} = 0 \quad$ at $\quad \varphi = \varphi_e$.
(2) $\mathscr{W} = 0, \quad M_{11} = 0 \quad$ at $\quad \varphi = \varphi_e$.

Determine the maximum values of N_{22} and M_{11} for the two types of support conditions.

4

THE THEORY OF SHELLS IN TENSOR NOTATION

4.1 Introduction

Tensor calculus is widely used in the modern treatment of the theory of shells (see, for example, references 18, 20, 22, and 28). Formulating the governing equations of the theory of shells in tensor notation offers several important advantages, among which the generality and the compact form particularly can be emphasized. The tensor equations are valid in any curvilinear coordinate system on the middle surface of the shell, and the governing equations of any given shell can therefore be obtained by choosing a convenient system of curvilinear coordinates on the middle surface, and then specializing the general tensor equations for this particular system of surface coordinates. It follows that when tensor methods are used, the determination of the lines of curvature can be avoided (it will be recalled that a knowledge of the lines of curvature was a prerequisite for use of the equations of Chapter 1). Further, in many cases, the tensor notation makes it possible to write long and complicated formulae in a simple and compact form, thus simplifying theoretical derivations.

Because of these advantages of tensor methods, this final chapter will be devoted to a derivation of the governing equations of the theory of shells in tensor form. We shall, in principle, use the same methods of derivation as those employed in Chapter 1, but the details of the derivations will differ to some extent from those of that chapter because of the use of arbitrary coordinate curves on the middle surface instead of the lines of curvature.

In the following, it will be assumed that the reader is familiar with the elements of tensor calculus (corresponding approximately to the contents of Chapter 1 in the book by Green and Zerna[20]). More detailed treatment of tensor calculus can be found in references 39, 42, and 43.

4.2 The theory of surfaces in tensor form

We begin by summarizing some results of the theory of surfaces in tensor notation. This will give us the opportunity of explaining the notation, and we shall also be able to refer to the relevant formulae of the theory of surfaces in connection with the derivations in the following sections.

In the following, *greek indices* have the range 1, 2, and the *summation convention* is employed. A *vector* in three-dimensional space is denoted by a bold-faced letter and is determined by the three components of the vector in a fixed, right-handed, Cartesian coordinate system.

The middle surface of the shell is determined by the parametric representation (cf. (1.1)):

$$\mathbf{r} = \mathbf{r}(\theta^1, \theta^2). \tag{4.1}$$

This representation defines a system of curvilinear coordinates on the surface. While it was assumed in the greater part of Chapter 1 that the coordinate curves were the lines of curvature, the present derivation is valid for arbitrary coordinate curves.

The *covariant base vectors* of the middle surface are defined by

$$\mathbf{a}_\alpha = \mathbf{r}_{,\alpha}, \tag{4.2}$$

the *comma notation* being used to denote partial derivatives with respect to θ^α. It will be assumed that the condition

$$|\mathbf{a}_1 \times \mathbf{a}_2| > 0 \tag{4.2a}$$

is satisfied everywhere on the middle surface. In any transformation of curvilinear coordinates on the middle surface, we find for the base vectors $\bar{\mathbf{a}}_\alpha$ associated with the new surface coordinates $\bar{\theta}^\alpha$:

$$\bar{\mathbf{a}}_\alpha = \frac{\partial \mathbf{r}}{\partial \bar{\theta}^\alpha} = \frac{\partial \mathbf{r}}{\partial \theta^\lambda}\frac{\partial \theta^\lambda}{\partial \bar{\theta}^\alpha} = \mathbf{r}_{,\lambda}\frac{\partial \theta^\lambda}{\partial \bar{\theta}^\alpha} = \frac{\partial \theta^\lambda}{\partial \bar{\theta}^\alpha}\mathbf{a}_\lambda. \tag{4.3}$$

The vectors $\mathbf{a}_\alpha$ therefore transform according to the rule for covariant tensors of order one (i.e. each of the Cartesian components of the vectors $\mathbf{a}_\alpha$ transforms according to the said rule), and this is the motivation for the name 'covariant base vectors'.

In the following we shall confine ourselves to transformations of surface coordinates for which

$$\det(\partial\theta^\lambda/\partial\bar{\theta}^\alpha) > 0,$$

i.e. the Jacobian is positive everywhere on the middle surface. This condition ensures that the orientation of the curvilinear coordinate system on the middle surface (i.e. the sense of rotation determined by the surface coordinate system) is preserved in the transformation.

The *metric tensors* of the surface are defined by

$$a_{\alpha\beta} = \mathbf{a}_\alpha \cdot \mathbf{a}_\beta = a_{\beta\alpha}, \qquad a^{\alpha\lambda}a_{\lambda\beta} = \delta^\alpha_\beta. \tag{4.4}$$

$a_{\alpha\beta}$ and $a^{\alpha\beta}$ are covariant and contravariant symmetric tensors of order two, respectively.

The length of an infinitesimal line element on the surface connecting two points with curvilinear coordinates θ^α and $\theta^\alpha + d\theta^\alpha$ is determined by

$$(ds)^2 = a_{\alpha\beta}\, d\theta^\alpha\, d\theta^\beta, \tag{4.5}$$

the right-hand side of which is the *first fundamental form* of the surface (cf. (1.10)). Let

$$a = \det(a_{\alpha\beta}) \tag{4.6}$$

denote the determinant associated with the metric tensor. The *element of surface area* is then given by

$$dA = (\sqrt{a})\, d\theta^1\, d\theta^2. \tag{4.7}$$

It follows from the relation $|\mathbf{a}_1 \times \mathbf{a}_2|^2 = a$ and the assumption (4.2a) that $a > 0$ everywhere on the middle surface.

The *contravariant base vectors* of the surface are given by

$$\mathbf{a}^\alpha = a^{\alpha\beta}\mathbf{a}_\beta, \tag{4.8}$$

and we have the formulae

$$\mathbf{a}^\alpha \cdot \mathbf{a}^\beta = a^{\alpha\beta}, \qquad \mathbf{a}^\alpha \cdot \mathbf{a}_\beta = \delta^\alpha_\beta. \tag{4.9}$$

From (4.8) and the fact that $\mathbf{a}_\beta$ transform according to the rule for covariant tensors of order one, it follows that the vectors $\mathbf{a}^\alpha$ transform according to the rule for contravariant tensors of order one.

The *unit normal vector* of the surface is given by

$$\mathbf{a}_3 = \frac{1}{\sqrt{a}}\mathbf{a}_1 \times \mathbf{a}_2, \tag{4.10}$$

and we have the relations

$$\mathbf{a}_3 \cdot \mathbf{a}_\alpha = 0, \qquad \mathbf{a}_3 \cdot \mathbf{a}^\alpha = 0. \tag{4.11}$$

The *permutation tensors* are denoted by $e_{\alpha\beta}$ and $e^{\alpha\beta}$ and are defined by

$$\begin{aligned} e_{\alpha\beta}&: \quad e_{12} = -e_{21} = \sqrt{a}, \qquad e_{11} = e_{22} = 0, \\ e^{\alpha\beta}&: \quad e^{12} = -e^{21} = 1/\sqrt{a}, \qquad e^{11} = e^{22} = 0, \end{aligned} \tag{4.12}$$

and we have the formulae

$$\begin{aligned} e_{\alpha\beta} &= -e_{\beta\alpha}, & e^{\alpha\beta} &= -e^{\beta\alpha}, \\ e^{\alpha\lambda}e_{\beta\lambda} = e^{\lambda\alpha}e_{\lambda\beta} &= \delta^\alpha_\beta, & e^{\alpha\lambda}e_{\beta\mu} &= \delta^\alpha_\beta\delta^\lambda_\mu - \delta^\alpha_\mu\delta^\lambda_\beta. \end{aligned} \tag{4.13}$$

The following expressions are obtained for the vector products of the base vectors:

$$\begin{aligned} \mathbf{a}_\alpha \times \mathbf{a}_\beta &= e_{\alpha\beta}\mathbf{a}_3, & \mathbf{a}^\alpha \times \mathbf{a}^\beta &= e^{\alpha\beta}\mathbf{a}_3, \\ \mathbf{a}_3 \times \mathbf{a}_\alpha &= e_{\alpha\beta}\mathbf{a}^\beta, & \mathbf{a}_3 \times \mathbf{a}^\alpha &= e^{\alpha\beta}\mathbf{a}_\beta. \end{aligned} \tag{4.14}$$

We also introduce the *transverse vectors* associated with the base vectors (these are obtained by rotating the base vectors by an angle $\pi/2$ in the tangent plane (see Fig. 8, section 1.4.3)). These transverse vectors are given by

$$\hat{\mathbf{a}}_\alpha = \mathbf{a}_3 \times \mathbf{a}_\alpha = e_{\alpha\beta}\mathbf{a}^\beta, \qquad \hat{\mathbf{a}}^\alpha = \mathbf{a}_3 \times \mathbf{a}^\alpha = e^{\alpha\beta}\mathbf{a}_\beta, \tag{4.15}$$

and we have

$$\begin{gathered} \hat{\mathbf{a}}_\alpha \cdot \hat{\mathbf{a}}_\beta = a_{\alpha\beta}, \qquad \hat{\mathbf{a}}^\alpha \cdot \hat{\mathbf{a}}^\beta = a^{\alpha\beta}, \qquad \hat{\mathbf{a}}^\alpha \cdot \hat{\mathbf{a}}_\beta = \delta^\alpha_\beta, \\ \mathbf{a}_\alpha = e_{\beta\alpha}\hat{\mathbf{a}}^\beta, \qquad \mathbf{a}^\alpha = e^{\beta\alpha}\hat{\mathbf{a}}_\beta, \qquad \mathbf{a}^\alpha \times \hat{\mathbf{a}}_\beta = -\hat{\mathbf{a}}^\alpha \times \mathbf{a}_\beta = \delta^\alpha_\beta \mathbf{a}_3. \end{gathered} \tag{4.16}$$

Noting that the unit normal vector $\mathbf{a}_3$ is an invariant, and that the base vectors $\mathbf{a}_\alpha$ and $\mathbf{a}^\alpha$ transform according to the rules for covariant and contravariant tensors of order one, respectively, we conclude from the intermediate expressions in the formulae (4.15) that the transverse vectors $\hat{\mathbf{a}}_\alpha$ and $\hat{\mathbf{a}}^\alpha$ transform according to the rules for covariant and contravariant tensors of order one, respectively.

The coefficients of the *second fundamental form* of the surface are given by (cf. (1.18))

$$b_{\alpha\beta} = b_{\beta\alpha} = \mathbf{a}_{\alpha,\beta} \cdot \mathbf{a}_3 = -\mathbf{a}_{3,\alpha} \cdot \mathbf{a}_\beta. \tag{4.17}$$

$b_{\alpha\beta}$ is a symmetric, covariant tensor of order two.

The *derivatives of the base vectors* are determined by

$$\begin{gathered} \mathbf{a}_{\alpha,\beta} = \mathbf{a}_{\beta,\alpha} = \begin{Bmatrix} \lambda \\ \alpha \quad \beta \end{Bmatrix} \mathbf{a}_\lambda + b_{\alpha\beta}\mathbf{a}_3, \\ \mathbf{a}^\alpha{}_{,\beta} = -\begin{Bmatrix} \alpha \\ \lambda \quad \beta \end{Bmatrix} \mathbf{a}^\lambda + b^\alpha_\beta \mathbf{a}_3, \qquad \mathbf{a}_{3,\alpha} = -b^\beta_\alpha \mathbf{a}_\beta. \end{gathered} \tag{4.18}$$

The *Christoffel symbols* of the second kind for the surface are defined by

$$\begin{Bmatrix} \gamma \\ \alpha \quad \beta \end{Bmatrix} = \mathbf{a}_{\alpha,\beta} \cdot \mathbf{a}^\gamma = -\mathbf{a}^\gamma{}_{,\beta} \cdot \mathbf{a}_\alpha = -\mathbf{a}^\gamma{}_{,\alpha} \cdot \mathbf{a}_\beta. \tag{4.19}$$

The Christoffel symbols can be expressed in the following manner in terms of the metric tensor

$$\begin{Bmatrix} \gamma \\ \alpha \quad \beta \end{Bmatrix} = a^{\gamma\lambda}\tfrac{1}{2}(a_{\beta\lambda,\alpha} + a_{\lambda\alpha,\beta} - a_{\alpha\beta,\lambda}), \tag{4.20}$$

and we have the formula

$$\begin{Bmatrix} \lambda \\ \lambda \quad \alpha \end{Bmatrix} = \frac{1}{\sqrt{a}} \frac{\partial \sqrt{a}}{\partial \theta^\alpha}. \tag{4.21}$$

It should be noted that the Christoffel symbols are not, in general, the components of a tensor.

Covariant derivatives are denoted by vertical lines. Thus, in the case of a contravariant tensor f^α of order one, we have

$$f^\alpha{}_{|\beta} = f^\alpha{}_{,\beta} + \begin{Bmatrix} \alpha \\ \lambda \quad \beta \end{Bmatrix} f^\lambda. \tag{4.22}$$

The covariant derivative $f^{\alpha}{}_{|\beta}$ is a mixed tensor of order two. It can be shown that the covariant derivatives of the metric tensors $a_{\alpha\beta}$, $a^{\alpha\beta}$ and of the permutation tensors $e_{\alpha\beta}$, $e^{\alpha\beta}$ are identically zero.

The *covariant derivatives of the base vectors* are defined by

$$\mathbf{a}_{\alpha|\beta} = \mathbf{a}_{\alpha,\beta} - \begin{Bmatrix} \lambda \\ \alpha \quad \beta \end{Bmatrix} \mathbf{a}_{\lambda} = b_{\alpha\beta}\mathbf{a}_3. \tag{4.23a}$$

As $b_{\alpha\beta}$ is a covariant tensor and $\mathbf{a}_3$ is an invariant, it follows from (4.23a) that $\mathbf{a}_{\alpha|\beta}$ transforms according to the rule for covariant tensors of order two.

In a similar manner we find

$$\mathbf{a}^{\alpha}{}_{|\beta} = b^{\alpha}_{\beta}\mathbf{a}_3, \qquad \mathbf{a}_{3|\alpha} = -b^{\beta}_{\alpha}\mathbf{a}_{\beta}. \tag{4.23b}$$

In the latter equation we have used the fact that the covariant derivative of the invariant $\mathbf{a}_3$ equals the partial derivative of this quantity.

For the covariant derivatives of the transverse vectors we find

$$\begin{aligned} \hat{\mathbf{a}}_{\alpha|\beta} &= (\mathbf{a}_3 \times \mathbf{a}_{\alpha})_{|\beta} = \mathbf{a}_{3,\beta} \times \mathbf{a}_{\alpha} = e_{\alpha\lambda}b^{\lambda}_{\beta}\mathbf{a}_3, \\ \hat{\mathbf{a}}^{\alpha}{}_{|\beta} &= e^{\alpha\lambda}b_{\lambda\beta}\mathbf{a}_3, \qquad \mathbf{a}_{3,\alpha} = -b^{\lambda}_{\alpha}e_{\beta\lambda}\hat{\mathbf{a}}^{\beta}. \end{aligned} \tag{4.24}$$

The *Codazzi equations* of the surface can be written in the form

$$b_{\alpha\beta|\gamma} = b_{\alpha\gamma|\beta}. \tag{4.25}$$

Only two of these equations are independent, namely the following:

$$b_{11|2} = b_{12|1}, \qquad b_{22|1} = b_{21|2}. \tag{4.25a}$$

The *Riemann–Christoffel tensor* of the surface is defined by

$$R^{\alpha}_{\cdot\,\beta\gamma\delta} = \begin{Bmatrix} \alpha \\ \beta \quad \delta \end{Bmatrix}_{,\gamma} - \begin{Bmatrix} \alpha \\ \beta \quad \gamma \end{Bmatrix}_{,\delta} + \begin{Bmatrix} \lambda \\ \beta \quad \delta \end{Bmatrix}\begin{Bmatrix} \alpha \\ \lambda \quad \gamma \end{Bmatrix} - \begin{Bmatrix} \lambda \\ \beta \quad \gamma \end{Bmatrix}\begin{Bmatrix} \alpha \\ \lambda \quad \delta \end{Bmatrix}. \tag{4.26}$$

For the associated tensor $R_{\alpha\beta\gamma\delta}$ we have the relations

$$R_{1212} = R_{2121} = -R_{1221} = -R_{2112}, \tag{4.27}$$

while all the other components of $R_{\alpha\beta\gamma\delta}$ are identically zero. We also have the following relation between the Riemann–Christoffel tensor and the coefficients of the second fundamental form:

$$R_{\alpha\beta\gamma\delta} = b_{\alpha\gamma}b_{\beta\delta} - b_{\alpha\delta}b_{\beta\gamma}. \tag{4.28}$$

Only one of the components of (4.28) gives rise to a non-trivial result, namely the following:

$$R_{1212} = b_{11}b_{22} - (b_{12})^2. \tag{4.28a}$$

This is the *Gauss equation* of the surface.

In the case of repeated covariant differentiations, the value of the result will, in general, depend on the order of differentiations. If we consider the second covariant derivatives of a covariant vector, it is found that $v_{\alpha|\beta\gamma}$ will, in general, be different from $v_{\alpha|\gamma\beta}$, and we have the relation

$$v_{\alpha|\beta\gamma} - v_{\alpha|\gamma\beta} = R^{\lambda}_{.\alpha\beta\gamma} v_{\lambda}. \tag{4.29}$$

Green's theorem. Let f^{α} be a contravariant vector of order one defined on the middle surface. Let C be the boundary curve of the shell, let $\mathbf{t}$ be a unit tangent vector, and let $\boldsymbol{\nu}$ be an outward unit normal vector to the boundary curve. We then have the relations

$$\begin{aligned} \mathbf{t} &= t^{\alpha}\mathbf{a}_{\alpha} = \frac{\mathrm{d}\theta^{\alpha}}{\mathrm{d}s}\,\mathbf{a}_{\alpha}, \\ \boldsymbol{\nu} &= \nu_{\alpha}\mathbf{a}^{\alpha} = e_{\alpha\beta}t^{\beta}\mathbf{a}^{\alpha}, \end{aligned} \tag{4.30}$$

and Green's theorem assumes the form (cf. (1.113))

$$\int f^{\alpha}{}_{|\alpha}\,\mathrm{d}A = \oint f^{\alpha}\nu_{\alpha}\,\mathrm{d}s, \tag{4.31}$$

in which the element of surface area $\mathrm{d}A$ is given by (4.7).

Derivatives of vectors. We now consider a vector field on the middle surface, which associates a vector $\mathbf{v}(\theta^1, \theta^2)$ with each point of the surface. The vector $\mathbf{v}$ is resolved in the directions of the corresponding base vectors $\mathbf{a}^{\alpha}$ and normal vector $\mathbf{a}_3$. Thus

$$\mathbf{v} = v_{\beta}\mathbf{a}^{\beta} + v_3\mathbf{a}_3. \tag{4.32}$$

We now take the scalar product of this equation with $\mathbf{a}_{\alpha}$ and $\mathbf{a}_3$, respectively, to obtain the formulae

$$v_{\alpha} = \mathbf{v}\cdot\mathbf{a}_{\alpha}, \qquad v_3 = \mathbf{v}\cdot\mathbf{a}_3. \tag{4.33}$$

As $\mathbf{v}$ and $\mathbf{a}_3$ are invariants for transformations of surface coordinates, while the vectors $\mathbf{a}_{\alpha}$ transform according to the rule for covariant tensors of order one, it follows from (4.33) that v_{α} is a covariant tensor of order one, and v_3 is an invariant.

By using the formulae (4.23), we find for the derivatives of the vector $\mathbf{v}$:

$$\mathbf{v}_{,\alpha} = \mathbf{v}_{|\alpha} = (v_{\beta|\alpha} - b_{\alpha\beta}v_3)\mathbf{a}^{\beta} + (v_{3,\alpha} + b^{\beta}_{\alpha}v_{\beta})\mathbf{a}_3. \tag{4.34}$$

In the following we shall occasionally wish to resolve a vector in the directions of the transverse vectors $\hat{\mathbf{a}}^{\alpha}$ and the normal vector $\mathbf{a}_3$. We therefore consider a vector $\boldsymbol{\omega}(\theta^1, \theta^2)$ represented in the form

$$\boldsymbol{\omega} = \omega_{\beta}\hat{\mathbf{a}}^{\beta} + \omega_3\mathbf{a}_3, \tag{4.35}$$

An argument similar to the one following equation (4.33) enables us to deduce that ω_{β} is a covariant tensor of order one, and ω_3 is an invariant. Using the formulae (4.24) we then find

$$\boldsymbol{\omega}_{,\alpha} = \boldsymbol{\omega}_{|\alpha} = (\omega_{\beta|\alpha} - e_{\beta\lambda}b^{\lambda}_{\alpha}\omega_3)\hat{\mathbf{a}}^{\beta} + (\omega_{3,\alpha} + e^{\beta\lambda}b_{\lambda\alpha}\omega_{\beta})\mathbf{a}_3. \tag{4.36}$$

For the purpose of later applications, we shall also consider the following case. Let two vectors $\mathbf{N}^{\alpha}$ ($\alpha = 1, 2$) be associated with each point of the middle surface, and let it be given that the vectors $\mathbf{N}^\alpha$ transform according to the rule for contravariant tensors of order one for transformations of surface coordinates. We resolve $\mathbf{N}^\alpha$ in the directions of $\mathbf{a}_\alpha$ and $\mathbf{a}_3$ and write

$$\mathbf{N}^\alpha = N^{\alpha\beta}\mathbf{a}_\beta + Q^\alpha\mathbf{a}_3. \tag{4.37}$$

An argument similar to the one following (4.33) enables us to deduce that $N^{\alpha\beta}$ and Q^α are tensors, the types and orders of which are given by the positions and the numbers of the indices. We now form the covariant derivative of (4.37) and use (4.23) to obtain

$$\mathbf{N}^\alpha{}_{|\lambda} = (N^{\alpha\beta}{}_{|\lambda} - b^\beta_\lambda Q^\alpha)\mathbf{a}_\beta + (Q^\alpha{}_{|\lambda} + N^{\alpha\beta}b_{\beta\lambda})\mathbf{a}_3. \tag{4.38}$$

Next, we consider the case in which the two vectors $\mathbf{M}^\alpha$ at each point of the middle surface are situated on the tangent plane at the point, the vectors $\mathbf{M}^\alpha$ transforming according to the rule for contravariant tensors of order one. We resolve $\mathbf{M}^\alpha$ in the directions of the transverse vectors $\hat{\mathbf{a}}_\beta$, i.e. we put

$$\mathbf{M}^\alpha = M^{\alpha\beta}\hat{\mathbf{a}}_\beta. \tag{4.39}$$

By the previous argument, $M^{\alpha\beta}$ is seen to be a contravariant tensor of order two. Covariant differentiation of (4.39) followed by application of (4.24) then gives

$$\mathbf{M}^\alpha{}_{|\lambda} = M^{\alpha\beta}{}_{|\lambda}\hat{\mathbf{a}}_\beta - e_{\mu\beta}b^\mu_\lambda M^{\alpha\beta}\mathbf{a}_3. \tag{4.40}$$

We shall finally determine the covariant derivative of vectors of the type $\mathbf{k}_\alpha$, which transform according to the rule for covariant tensors of order one. We put

$$\mathbf{k}_\alpha = k_{\alpha\gamma}\hat{\mathbf{a}}^\gamma + k_\alpha\mathbf{a}_3, \tag{4.41}$$

which, by the previous argument, implies that $k_{\alpha\gamma}$ and k_α are tensors, the types and orders of which are given by the positions and the numbers of the indices. Covariant differentiation of (4.41) followed by application of (4.24) then yields

$$\mathbf{k}_{\alpha|\beta} = (k_{\alpha\mu|\beta} - e_{\mu\lambda}b^\lambda_\beta k_\alpha)\hat{\mathbf{a}}^\mu + (k_{\alpha|\beta} + e^{\lambda\mu}b_{\mu\beta}k_{\alpha\lambda})\mathbf{a}_3. \tag{4.42}$$

4.3 Geometrical description

In the following derivations we shall retain the assumptions introduced in Chapter 1, namely that *the displacements and the rotations* of the middle surface *are infinitesimally small*.

4.3.1 The middle surface strains

The displacement vector $\mathbf{v}(\theta^1, \theta^2)$ is resolved in the directions of the base vectors $\mathbf{a}^\alpha$ and the normal vector $\mathbf{a}_3$, so that

$$\mathbf{v} = v_\beta\mathbf{a}^\beta + v_3\mathbf{a}_3. \tag{4.43}$$

In the present case, the strain measures of the middle surface will be defined by (cf. (1.61))

$$\varepsilon_{\alpha\beta} = \tfrac{1}{2}(a^*_{\alpha\beta} - a_{\alpha\beta})_{\text{lin}} = \tfrac{1}{2}\delta a_{\alpha\beta}. \tag{4.44}$$

As the difference $(a^*_{\alpha\beta} - a_{\alpha\beta})$ between the metric tensors in the deformed state and in the reference state is again a tensor, it follows from formula (4.44) that $\varepsilon_{\alpha\beta}$ is a symmetric, covariant tensor of order two (the *strain tensor* of the middle surface). Inserting (1.57) into (4.44) we obtain

$$\varepsilon_{\alpha\beta} = \tfrac{1}{2}(\mathbf{a}_\alpha \cdot \mathbf{v}_{,\beta} + \mathbf{a}_\beta \cdot \mathbf{v}_{,\alpha}). \tag{4.45}$$

Application of (4.34) then gives

$$\varepsilon_{\alpha\beta} = \tfrac{1}{2}(v_{\alpha|\beta} + v_{\beta|\alpha}) - b_{\alpha\beta}v_3, \tag{4.46}$$

which expresses the strain tensor in terms of the displacement components.

4.3.2 The rotation vector

Using formulae (4.45), $(4.13)_4$, $(4.14)_3$, and $(4.16)_6$, we transform the derivative of the displacement vector $\mathbf{v}_{,\alpha}$ in the following manner:

$$\begin{aligned}\mathbf{v}_{,\alpha} &= (\mathbf{v}_{,\alpha} \cdot \mathbf{a}_\beta)\mathbf{a}^\beta + (\mathbf{v}_{,\alpha} \cdot \mathbf{a}_3)\mathbf{a}_3 \\ &= \tfrac{1}{2}(\mathbf{v}_{,\alpha} \cdot \mathbf{a}_\beta + \mathbf{v}_{,\beta} \cdot \mathbf{a}_\alpha)\mathbf{a}^\beta + \tfrac{1}{2}(\mathbf{v}_{,\alpha} \cdot \mathbf{a}_\beta - \mathbf{v}_{,\beta} \cdot \mathbf{a}_\alpha)\mathbf{a}^\beta + (\mathbf{v}_{,\alpha} \cdot \mathbf{a}_3)\mathbf{a}_3 \\ &= \varepsilon_{\alpha\beta}\mathbf{a}^\beta + \tfrac{1}{2}e^{\lambda\mu}\mathbf{v}_{,\lambda} \cdot \mathbf{a}_\mu e_{\alpha\beta}\mathbf{a}^\beta - (\mathbf{v}_{,\beta} \cdot \mathbf{a}_3)\hat{\mathbf{a}}^\beta \times \mathbf{a}_\alpha \\ &= \varepsilon_{\alpha\beta}\mathbf{a}^\beta + [(\tfrac{1}{2}e^{\lambda\mu}\mathbf{v}_{,\lambda} \cdot \mathbf{a}_\mu)\mathbf{a}_3 - (\mathbf{v}_{,\beta} \cdot \mathbf{a}_3)\hat{\mathbf{a}}^\beta] \times \mathbf{a}_\alpha.\end{aligned} \tag{4.47}$$

Denoting the vector in brackets by $\boldsymbol{\omega}$, we obtain the formula (cf. (1.75))

$$\mathbf{v}_{,\alpha} = \varepsilon_{\alpha\beta}\mathbf{a}^\beta + \boldsymbol{\omega} \times \mathbf{a}_\alpha, \tag{4.48}$$

where

$$\boldsymbol{\omega} = \omega_\alpha \hat{\mathbf{a}}^\alpha + \omega_3 \mathbf{a}_3, \tag{4.49a}$$

and

$$\omega_\alpha = -\mathbf{a}_3 \cdot \mathbf{v}_{,\alpha}, \qquad \omega_3 = \tfrac{1}{2}e^{\alpha\beta}\mathbf{v}_{,\alpha} \cdot \mathbf{a}_\beta. \tag{4.49b}$$

As $\mathbf{a}_3$ is an invariant, and $\mathbf{v}_{,\alpha}$ transforms according to the rule for covariant tensors of order one, it follows from the right-hand sides of (4.49b) that ω_α is a covariant tensor of order one, and that ω_3 is an invariant. Formula (4.49a) then shows that $\boldsymbol{\omega}$ is an invariant.

The vector $\boldsymbol{\omega}$ is, in fact, identical with the *rotation vector* introduced in section 1.4.3. This may be shown in the following manner. From (4.49a and b) we obtain

$$\boldsymbol{\omega} \cdot \hat{\mathbf{a}}_\alpha = \omega_\alpha = -\mathbf{a}_3 \cdot \mathbf{v}_{,\alpha},$$

which is identical with (1.81). If the lines of curvature are used as coordinate curves, we further find that (4.49b) is identical with (1.82), since $\sqrt{a} = G = A_1 A_2$ in this special coordinate system. We conclude that the two vectors are identical since they have the same coordinates in the special coordinate system.

The formula (1.74) may be used without change in the present context, since the vectors appearing in this formula are all invariants. Thus

$$\delta \mathbf{a}_3 = \boldsymbol{\omega} \times \mathbf{a}_3. \tag{4.50}$$

Taking the scalar products of (4.48) with the vectors $\mathbf{a}_\beta$ and $\mathbf{a}_3$, respectively, we get

$$\begin{aligned} (\mathbf{v}_{,\alpha} - \boldsymbol{\omega} \times \mathbf{a}_\alpha) \cdot \mathbf{a}_\beta &= \varepsilon_{\alpha\beta}, \\ (\mathbf{v}_{,\alpha} - \boldsymbol{\omega} \times \mathbf{a}_\alpha) \cdot \mathbf{a}_3 &= 0. \end{aligned} \tag{4.51}$$

Inserting (4.34) in (4.49b), we finally obtain

$$\omega_\alpha = -(v_{3,\alpha} + b_\alpha^\beta v_\beta), \qquad \omega_3 = \tfrac{1}{2} e^{\alpha\beta} v_{\beta|\alpha}, \tag{4.52}$$

which gives the components of the rotation vector in terms of the displacements.

4.3.3 *The bending of the middle surface*

Proceeding in the same manner as in section 1.4.4, we begin with a calculation of the first variation of the coefficients of the second fundamental form. Using the formulae (4.17), (4.50), (1.55), and (4.23), we find

$$\begin{aligned} \delta b_{\alpha\beta} &= -\delta(\mathbf{a}_{3,\alpha} \cdot \mathbf{a}_\beta) \\ &= -\delta \mathbf{a}_{3,\alpha} \cdot \mathbf{a}_\beta - \mathbf{a}_{3,\alpha} \cdot \delta \mathbf{a}_\beta \\ &= -(\boldsymbol{\omega} \times \mathbf{a}_3)_{,\alpha} \cdot \mathbf{a}_\beta + b_\alpha^\lambda \mathbf{a}_\lambda \cdot \mathbf{v}_{,\beta} \\ &= -\boldsymbol{\omega}_{,\alpha} \cdot \hat{\mathbf{a}}_\beta + (\boldsymbol{\omega} \times b_\alpha^\lambda \mathbf{a}_\lambda) \cdot \mathbf{a}_\beta + b_\alpha^\lambda \mathbf{a}_\lambda \cdot \mathbf{v}_{,\beta} \\ &= -\boldsymbol{\omega}_{,\alpha} \cdot \hat{\mathbf{a}}_\beta + b_\alpha^\lambda \mathbf{a}_\lambda \cdot (\mathbf{v}_{,\beta} - \boldsymbol{\omega} \times \mathbf{a}_\beta). \end{aligned} \tag{4.53}$$

Using $(4.51)_1$, equation (4.53) can be written in the form

$$\delta b_{\alpha\beta} = -k_{\alpha\beta} + b_\alpha^\lambda \varepsilon_{\lambda\beta}, \tag{4.54}$$

where

$$k_{\alpha\beta} = \boldsymbol{\omega}_{,\alpha} \cdot \hat{\mathbf{a}}_\beta \tag{4.55}$$

(cf. (1.84) and (1.85)). As $\hat{\mathbf{a}}_\beta$ and $\boldsymbol{\omega}_{,\alpha}$ both transform according to the rule for covariant tensors of order one, it follows from the right-hand side of (4.55) that $k_{\alpha\beta}$ is a covariant tensor of order two. By using (4.36) and (4.52), we find

$$\begin{aligned} k_{\alpha\beta} &= \omega_{\beta|\alpha} - e_{\beta\lambda} b_\alpha^\lambda \omega_3 \\ &= -[v_{3|\beta\alpha} + b_{\beta|\alpha}^\lambda v_\lambda + b_\beta^\lambda v_{\lambda|\alpha} + \tfrac{1}{2} b_\alpha^\lambda (v_{\lambda|\beta} - v_{\beta|\lambda})]. \end{aligned} \tag{4.56}$$

We then obtain the following expressions for the *bending tensor* $\kappa_{\alpha\beta}$ (cf. (1.95) and (1.96)):

$$\begin{aligned}\kappa_{\alpha\beta} &= \tfrac{1}{2}(k_{\alpha\beta} + k_{\beta\alpha})\\ &= \tfrac{1}{2}(\boldsymbol{\omega}_{,\alpha} \cdot \hat{\mathbf{a}}_\beta + \boldsymbol{\omega}_{,\beta} \cdot \hat{\mathbf{a}}_\alpha)\\ &= -\left[v_{3|\alpha\beta} + b^{\lambda}_{\alpha|\beta} v_\lambda + \tfrac{3}{4}(b^{\lambda}_{\alpha} v_{\lambda|\beta} + b^{\lambda}_{\beta} v_{\lambda|\alpha}) - \tfrac{1}{4}(b^{\lambda}_{\alpha} v_{\beta|\lambda} + b^{\lambda}_{\beta} v_{\alpha|\lambda})\right],\end{aligned} \tag{4.57}$$

since

$$v_{3|\alpha\beta} = v_{3,\alpha\beta} - \begin{Bmatrix} \lambda \\ \alpha \quad \beta \end{Bmatrix} v_{3,\lambda} = v_{3|\beta\alpha},$$

and

$$b^{\lambda}_{\alpha|\beta} = b^{\lambda}_{\beta|\alpha}$$

according to Codazzi's equations (4.25). The formula (4.57) expresses the bending tensor in terms of the displacement components.

4.3.4 *Equations of compatibility*

In the same manner as in section 1.4.5, we now assume that the strain and the bending tensors are given functions, and we seek to determine the displacement $\mathbf{v}$. Since $\delta b_{\alpha\beta}$ is symmetric in the indices α and β, we find from (4.54)

$$e^{\lambda\mu} k_{\lambda\mu} = e^{\lambda\mu} b^{\gamma}_{\lambda} \varepsilon_{\gamma\mu}. \tag{4.58}$$

Multiplying by $e_{\alpha\beta}$ and applying $(4.13)_4$ we get

$$k_{\alpha\beta} - k_{\beta\alpha} = 2k_{[\alpha\beta]} = b^{\gamma}_{\alpha}\varepsilon_{\gamma\beta} - b^{\gamma}_{\beta}\varepsilon_{\gamma\alpha}.$$

Using the relation $\kappa_{\alpha\beta} = k_{(\alpha\beta)}$, we further deduce that

$$k_{\alpha\beta} = k_{(\alpha\beta)} + k_{[\alpha\beta]} = \kappa_{\alpha\beta} + \tfrac{1}{2}(b^{\gamma}_{\alpha}\varepsilon_{\gamma\beta} - b^{\gamma}_{\beta}\varepsilon_{\gamma\alpha}), \tag{4.59}$$

so that $k_{\alpha\beta}$ is determined by $\kappa_{\alpha\beta}$ and $\varepsilon_{\alpha\beta}$.

The vector $\boldsymbol{\omega}_{,\alpha}$ is now written in the form

$$\boldsymbol{\omega}_{,\alpha} = k_{\alpha\beta}\hat{\mathbf{a}}^\beta + k_\alpha \mathbf{a}_3. \tag{4.60}$$

The components of $\boldsymbol{\omega}_{,\alpha}$ in the directions of $\hat{\mathbf{a}}^\beta$ are given by $\boldsymbol{\omega}_{,\alpha} \cdot \hat{\mathbf{a}}_\beta = k_{\alpha\beta}$ (see (4.55)), and we have also introduced the notation k_α for the components in the direction of $\mathbf{a}_3$. It was shown in section 4.3.3 that $k_{\alpha\beta}$ is a tensor. As $\boldsymbol{\omega}_{,\alpha}$ transforms according to the rule for covariant tensors of order one, and $k_\alpha = \boldsymbol{\omega}_{,\alpha} \cdot \mathbf{a}_3$, it follows that k_α is a covariant tensor of order one. By introducing the notation

$$\mathbf{k}_\alpha = k_{\alpha\beta}\hat{\mathbf{a}}^\beta + k_\alpha \mathbf{a}_3, \qquad \boldsymbol{\gamma}_\alpha = \varepsilon_{\alpha\beta}\mathbf{a}^\beta, \tag{4.61}$$

the equations (4.60) and (4.48) can be written in the form

$$\boldsymbol{\omega}_{,\alpha} = \mathbf{k}_\alpha, \qquad \mathbf{v}_{,\alpha} = \boldsymbol{\omega} \times \mathbf{a}_\alpha + \boldsymbol{\gamma}_\alpha. \tag{4.62}$$

Now when $\varepsilon_{\alpha\beta}$ and $\kappa_{\alpha\beta}$ (and therefore also $k_{\alpha\beta}$; see (4.59)) are known functions, the quantities $\boldsymbol{\gamma}_\alpha$ are also known, but the quantities $\mathbf{k}_\alpha$ are only partially known, because the components k_α have not yet been determined.

In order for the vectors $\boldsymbol{\omega}$ and $\mathbf{v}$ to be derivable from equations (4.62) by integration, it is necessary and sufficient that the following conditions be satisfied (cf. (1.101)):

$$\mathbf{k}_{1,2} = \mathbf{k}_{2,1}, \qquad (\boldsymbol{\omega} \times \mathbf{a}_1 + \boldsymbol{\gamma}_1)_{,2} = (\boldsymbol{\omega} \times \mathbf{a}_2 + \boldsymbol{\gamma}_2)_{,1}. \tag{4.63}$$

Now

$$\frac{1}{\sqrt{a}}(\mathbf{k}_{1,2} - \mathbf{k}_{2,1}) = e^{\alpha\beta}\mathbf{k}_{\alpha,\beta} = e^{\alpha\beta}\left(\mathbf{k}_{\alpha,\beta} - \begin{Bmatrix} \lambda \\ \alpha \quad \beta \end{Bmatrix}\mathbf{k}_\lambda\right) = e^{\alpha\beta}\mathbf{k}_{\alpha|\beta},$$

since $\begin{Bmatrix} \lambda \\ \alpha \quad \beta \end{Bmatrix}$ is symmetric in the indices α and β. The equations (4.63) can therefore be written in the invariant form

$$e^{\alpha\beta}\mathbf{k}_{\alpha|\beta} = \mathbf{0}, \qquad e^{\alpha\beta}(\boldsymbol{\omega} \times \mathbf{a}_\alpha + \boldsymbol{\gamma}_\alpha)_{|\beta} = \mathbf{0}. \tag{4.64}$$

If we perform the covariant differentiation in $(4.64)_2$ and use $(4.62)_1$ and the equation

$$\mathbf{a}_{\alpha|\beta} = \mathbf{a}_{\beta|\alpha},$$

we obtain

$$e^{\alpha\beta}(\mathbf{k}_\beta \times \mathbf{a}_\alpha + \boldsymbol{\gamma}_{\alpha|\beta}) = \mathbf{0}. \tag{4.65}$$

We now substitute the expressions (4.61) for $\mathbf{k}_\beta$ and $\boldsymbol{\gamma}_\alpha$ in (4.65). Using $(4.14)_3$, $(4.16)_6$, and (4.23b), we get

$$\begin{aligned} &e^{\alpha\beta}[(k_{\beta\lambda}\hat{\mathbf{a}}^\lambda + k_\beta\mathbf{a}_3) \times \mathbf{a}_\alpha + (\varepsilon_{\alpha\lambda}\mathbf{a}^\lambda)_{|\beta}] \\ &= e^{\alpha\beta}(-k_{\beta\alpha}\mathbf{a}_3 + k_\beta\hat{\mathbf{a}}_\alpha + \varepsilon_{\alpha\lambda|\beta}\mathbf{a}^\lambda + \varepsilon_{\alpha\lambda}b^\lambda_\beta\mathbf{a}_3) = \mathbf{0}. \end{aligned}$$

Taking the scalar product of this equation with $\mathbf{a}_3$ and $\mathbf{a}_\gamma$, respectively, we obtain (noting that $\mathbf{a}_\gamma \cdot \hat{\mathbf{a}}_\alpha = e_{\alpha\gamma}$; see (4.15)):

$$\begin{aligned} e^{\alpha\beta}k_{\alpha\beta} &= e^{\alpha\beta}b^\lambda_\alpha\varepsilon_{\lambda\beta}, \\ k_\gamma + e^{\alpha\beta}\varepsilon_{\alpha\gamma|\beta} &= 0, \end{aligned} \tag{4.66}$$

or

$$k_\alpha = e^{\lambda\mu}\varepsilon_{\mu\alpha|\lambda}. \tag{4.67}$$

Equation (4.66) is identical to (4.58), and (4.67) expresses k_α in terms of $\varepsilon_{\alpha\beta}$.

We now insert the expression (4.42) for $\mathbf{k}_{\alpha|\beta}$ in $(4.64)_1$. Taking the scalar products of the resulting equation with $\mathbf{a}_3$ and $\hat{\mathbf{a}}_\gamma$, respectively, we get

$$\begin{aligned} e^{\alpha\beta}(k_{\alpha|\beta} + e^{\lambda\mu}k_{\alpha\lambda}b_{\mu\beta}) &= 0, \\ e^{\alpha\beta}(k_{\alpha\gamma|\beta} - e_{\gamma\lambda}b^\lambda_\beta k_\alpha) &= 0. \end{aligned} \tag{4.68}$$

In these equations we shall substitute the expressions (4.67) and (4.59) for

k_α and $k_{\alpha\beta}$. We thus obtain from $(4.68)_1$

$$e^{\alpha\beta}e^{\lambda\mu}[\varepsilon_{\mu\alpha|\lambda\beta} + (\kappa_{\alpha\lambda} + k_{[\alpha\lambda]})b_{\mu\beta}] = 0,$$

or, by using the fact that $e^{\alpha\beta}e^{\lambda\mu}b_{\mu\beta}$ is symmetric in the indices α and λ, and by making a few changes of dummy indices,

$$e^{\alpha\lambda}e^{\beta\mu}(\kappa_{\alpha\beta}b_{\lambda\mu} - \varepsilon_{\alpha\beta|\lambda\mu}) = 0. \tag{4.69a}$$

Finally, from $(4.68)_2$ we obtain

$$e^{\alpha\beta}[\kappa_{\alpha\gamma|\beta} + \tfrac{1}{2}(b^\lambda_\alpha\varepsilon_{\lambda\gamma} - b^\lambda_\gamma\varepsilon_{\lambda\alpha})_{|\beta} + b^\lambda_\alpha(\varepsilon_{\beta\lambda|\gamma} - \varepsilon_{\beta\gamma|\lambda})] = 0. \tag{4.69b}$$

The equations (4.69) are the three *equations of compatibility* of the shell.

4.4 Statics of the shell

4.4.1 Applied loads, contact forces and couples

The distributed surface load $\mathbf{p}$ is resolved in the directions of the base vectors $\mathbf{a}_\alpha$ and the normal vector $\mathbf{a}_3$, i.e.

$$\mathbf{p} = p^\alpha\mathbf{a}_\alpha + p^3\mathbf{a}_3. \tag{4.70}$$

As $\mathbf{p}$ is an invariant, it follows that p^α is a contravariant tensor of order one and that p^3 is an invariant (see the remarks following equation (4.33)).

In the same way as in section 1.5.1, we shall use the symbols $\mathbf{n}$ and $\mathbf{m}$ to denote the contact force and the contact couple, respectively, measured per unit length of an arbitrary section curve on the middle surface. However, for sections along the coordinate curves, we shall make a slight change of notation compared with that in section 1.5.1. The contact force and couple acting on a small element of a section along a θ^2-curve corresponding to the increment $d\theta^2$ will now be denoted by

$$(\sqrt{a})\mathbf{N}^1\,d\theta^2, \qquad (\sqrt{a})\mathbf{M}^1\,d\theta^2.$$

Similarly, the contact force and couple acting on a small element of a section along a θ^1-curve will be denoted by

$$(\sqrt{a})\mathbf{N}^2\,d\theta^1, \qquad (\sqrt{a})\mathbf{M}^2\,d\theta^1.$$

In the same way as in section 1.5.1 it will be assumed that the contact couple vectors $\mathbf{M}^\alpha$ at any point of the middle surface lie in the corresponding tangent plane.

We now consider a small triangular area element of the middle surface bounded by an arbitrary section curve with unit normal vector $\boldsymbol{\nu}$ and by sections along coordinate curves (cf. Fig. 13 in section 1.5.1). By formulating the equilibrium conditions for the element, we obtain (cf. (1.107))

$$\mathbf{n} = (\sqrt{a})\left(\mathbf{N}^1\frac{d\theta^2}{ds} - \mathbf{N}^2\frac{d\theta^1}{ds}\right), \qquad \mathbf{m} = (\sqrt{a})\left(\mathbf{M}^1\frac{d\theta^2}{ds} - \mathbf{M}^2\frac{d\theta^1}{ds}\right). \tag{4.71}$$

With the help of (4.12) and (4.30), these equations can be written in the form

$$\mathbf{n} = e_{\alpha\beta}\mathbf{N}^{\alpha}t^{\beta}, \qquad \mathbf{m} = e_{\alpha\beta}\mathbf{M}^{\alpha}t^{\beta},$$

or

$$\mathbf{n} = \mathbf{N}^{\alpha}\nu_{\alpha}, \qquad \mathbf{m} = \mathbf{M}^{\alpha}\nu_{\alpha}. \tag{4.72}$$

Now, $\mathbf{n}$ and $\mathbf{m}$ are invariants for transformations of surface coordinates. Further, ν_{α} is a covariant tensor of order one, and the ratio between the components ν_1 and ν_2 can be chosen arbitrarily, since $\boldsymbol{\nu}$ is an arbitrary unit vector. It follows from the quotient law of tensors that $\mathbf{N}^{\alpha}$ and $\mathbf{M}^{\alpha}$ both transform according to the rule for contravariant tensors of order one. If we write $\mathbf{N}^{\alpha}$ and $\mathbf{M}^{\alpha}$ in the form

$$\mathbf{N}^{\alpha} = \mathrm{N}^{\alpha\beta}\mathbf{a}_{\beta} + \mathrm{Q}^{\alpha}\mathbf{a}_3, \qquad \mathbf{M}^{\alpha} = \mathrm{M}^{\alpha\beta}\hat{\mathbf{a}}_{\beta}, \tag{4.73}$$

it will be seen that $\mathrm{N}^{\alpha\beta}$, Q^{α}, and $\mathrm{M}^{\alpha\beta}$ are tensors, the types and orders of which are given by the positions and the numbers of indices (see section 4.2). The quantities $\mathrm{N}^{\alpha\beta}$, Q^{α}, and $\mathrm{M}^{\alpha\beta}$ are called the *complete contact forces and couples*.

4.4.2 *Equations of equilibrium*

If we formulate the equilibrium conditions for a small element of the middle surface, and consider the limiting case when the area of the element tends to zero, we obtain the equations (cf. Fig. 15 and (1.110))

$$((\sqrt{a})\mathbf{N}^{\alpha})_{,\alpha} + (\sqrt{a})\mathbf{p} = \mathbf{0}, \qquad ((\sqrt{a})\mathbf{M}^{\alpha})_{,\alpha} + \mathbf{a}_{\alpha} \times (\sqrt{a})\mathbf{N}^{\alpha} = \mathbf{0}. \tag{4.74}$$

Since

$$((\sqrt{a})\mathbf{N}^{\alpha})_{,\alpha} = (\sqrt{a})\mathbf{N}^{\alpha}{}_{,\alpha} + \frac{\partial\sqrt{a}}{\partial\theta^{\alpha}}\mathbf{N}^{\alpha} = (\sqrt{a})\left(\mathbf{N}^{\alpha}{}_{,\alpha} + \begin{Bmatrix} \lambda \\ \alpha \quad \lambda \end{Bmatrix}\mathbf{N}^{\alpha}\right) = (\sqrt{a})\mathbf{N}^{\alpha}{}_{|\alpha}$$

(see (4.21)), we see that (4.74) may be written in the form

$$\mathbf{N}^{\alpha}{}_{|\alpha} + \mathbf{p} = \mathbf{0}, \qquad \mathbf{M}^{\alpha}{}_{|\alpha} + \mathbf{a}_{\alpha} \times \mathbf{N}^{\alpha} = \mathbf{0}. \tag{4.75}$$

These are the *vector equations of equilibrium* of the shell. From (4.73), (4.14), and (4.15) we find the relations

$$\mathbf{a}_{\alpha} \times \mathbf{N}^{\alpha} = \mathbf{a}_{\alpha} \times (\mathrm{N}^{\alpha\beta}\mathbf{a}_{\beta} + \mathrm{Q}^{\alpha}\mathbf{a}_3) = \mathrm{N}^{\alpha\beta}e_{\alpha\beta}\mathbf{a}_3 - \mathrm{Q}^{\alpha}\hat{\mathbf{a}}_{\alpha}. \tag{4.76}$$

We now substitute (4.38), (4.40), and (4.76) into (4.75), and we then take the scalar products of $(4.75)_1$ with $\mathbf{a}^{\beta}$ and $\mathbf{a}_3$, respectively, and the scalar products of $(4.75)_2$ with $\hat{\mathbf{a}}^{\beta}$ and $\mathbf{a}_3$, respectively. Hence

$$\begin{aligned} &\mathrm{N}^{\alpha\beta}{}_{|\alpha} - b^{\beta}_{\alpha}\mathrm{Q}^{\alpha} + p^{\beta} = 0, \\ &\mathrm{Q}^{\alpha}{}_{|\alpha} + \mathrm{N}^{\alpha\beta}b_{\alpha\beta} + p^3 = 0, \\ &\mathrm{M}^{\alpha\beta}{}_{|\alpha} - \mathrm{Q}^{\beta} = 0, \\ &e_{\alpha\beta}(\mathrm{N}^{\alpha\beta} - b^{\alpha}_{\lambda}\mathrm{M}^{\lambda\beta}) = 0). \end{aligned} \tag{4.77}$$

These are the six *scalar equations of equilibrium* of the shell.

4.4.3 *The principle of virtual work*

In the same way as in section 1.5.3, we shall assume that the shell is in equilibrium, and we begin with the formula (1.112) for the external work. Substituting the expressions (4.72) for **n** and **m**, and transforming the resulting expression by means of Green's theorem (4.31), we obtain

$$\begin{aligned} A_{(e)} &= \int \mathbf{p} \cdot \mathbf{v}\, dA + \oint_C (\mathbf{n} \cdot \mathbf{v} + \mathbf{m} \cdot \boldsymbol{\omega})\, ds \\ &= \int \mathbf{p} \cdot \mathbf{v}\, dA + \oint_C (\mathbf{N}^\alpha \cdot \mathbf{v} + \mathbf{M}^\alpha \cdot \boldsymbol{\omega})\nu_\alpha\, ds \\ &= \int [\mathbf{p} \cdot \mathbf{v} + (\mathbf{N}^\alpha \cdot \mathbf{v})_{|\alpha} + (\mathbf{M}^\alpha \cdot \boldsymbol{\omega})_{|\alpha}]\, dA \\ &= \int [(\mathbf{N}^\alpha{}_{|\alpha} + \mathbf{p}) \cdot \mathbf{v} + \mathbf{M}^\alpha{}_{|\alpha} \cdot \boldsymbol{\omega} + \mathbf{N}^\alpha \cdot \mathbf{v}_{,\alpha} + \mathbf{M}^\alpha \cdot \boldsymbol{\omega}_{,\alpha}]\, dA. \end{aligned}$$

By assumption, the equations of equilibrium (4.75) are satisfied. Using these equations together with (4.73), (4.48), and (4.55), we deduce

$$\begin{aligned} A_{(e)} &= \int [\mathbf{N}^\alpha \cdot (\mathbf{v}_{,\alpha} - \boldsymbol{\omega} \times \mathbf{a}_\alpha) + \mathbf{M}^\alpha \cdot \boldsymbol{\omega}_{,\alpha}]\, dA \\ &= \int [(\mathrm{N}^{\alpha\beta}\mathbf{a}_\beta + \mathrm{Q}^\alpha \mathbf{a}_3) \cdot \varepsilon_{\alpha\lambda}\mathbf{a}^\lambda + \mathrm{M}^{\alpha\beta}\hat{\mathbf{a}}_\beta \cdot \boldsymbol{\omega}_{,\alpha}]\, dA \\ &= \int (\mathrm{N}^{\alpha\beta}\varepsilon_{\alpha\beta} + \mathrm{M}^{\alpha\beta}k_{\alpha\beta})\, dA. \end{aligned}$$

The last integral is called the internal work. The *principle of virtual work* can therefore be expressed in the form

$$\int \mathbf{p} \cdot \mathbf{v}\, dA + \oint_C (\mathbf{n} \cdot \mathbf{v} + \mathbf{m} \cdot \boldsymbol{\omega})\, ds = \int (\mathrm{N}^{\alpha\beta}\varepsilon_{\alpha\beta} + \mathrm{M}^{\alpha\beta}k_{\alpha\beta})\, dA. \quad (4.78)$$

If we write each of the tensors $\mathrm{N}^{\alpha\beta}$ and $\mathrm{M}^{\alpha\beta}$ as a sum of symmetric and skew-symmetric contributions (see (1.117)) and use the expression (4.59) for $k_{\alpha\beta}$, we find for the integrand on the right-hand side of (4.78) (noting that $\varepsilon_{\alpha\beta}$ and $\kappa_{\alpha\beta}$ are symmetric in the indices α and β):

$$\begin{aligned} \mathrm{N}^{\alpha\beta}\varepsilon_{\alpha\beta} + \mathrm{M}^{\alpha\beta}k_{\alpha\beta} &= \mathrm{N}^{(\alpha\beta)}\varepsilon_{\alpha\beta} + (\mathrm{M}^{(\alpha\beta)} + \mathrm{M}^{[\alpha\beta]})[\kappa_{\alpha\beta} + \tfrac{1}{2}(b^\gamma_\alpha\varepsilon_{\gamma\beta} - b^\gamma_\beta\varepsilon_{\gamma\alpha})] \\ &= \mathrm{N}^{(\alpha\beta)}\varepsilon_{\alpha\beta} + \mathrm{M}^{(\alpha\beta)}\kappa_{\alpha\beta} + \mathrm{M}^{[\alpha\beta]}b^\gamma_\alpha\varepsilon_{\gamma\beta} \\ &= [\mathrm{N}^{(\alpha\beta)} + \tfrac{1}{2}(b^\alpha_\lambda\mathrm{M}^{[\lambda\beta]} + b^\beta_\lambda\mathrm{M}^{[\lambda\alpha]})]\varepsilon_{\alpha\beta} + \mathrm{M}^{(\alpha\beta)}\kappa_{\alpha\beta}. \end{aligned}$$

If we define the *effective contact quantities* by

$$\begin{aligned} N^{\alpha\beta} &= N^{\beta\alpha} = \mathrm{N}^{(\alpha\beta)} + \tfrac{1}{2}(b^\alpha_\lambda\mathrm{M}^{[\lambda\beta]} + b^\beta_\lambda\mathrm{M}^{[\lambda\alpha]}), \\ M^{\alpha\beta} &= M^{\beta\alpha} = \mathrm{M}^{(\alpha\beta)}, \end{aligned} \quad (4.79)$$

the principle of virtual work (4.78) can be written in the form

$$\int \mathbf{p} \cdot \mathbf{v}\, dA + \oint_C (\mathbf{n} \cdot \mathbf{v} + \mathbf{m} \cdot \boldsymbol{\omega})\, ds = \int (N^{\alpha\beta}\varepsilon_{\alpha\beta} + M^{\alpha\beta}\kappa_{\alpha\beta})\, dA. \quad (4.80)$$

In a similar way as in section 1.5.3, we now introduce *effective contact forces and couples* defined by (cf. (1.121) and (1.122))

$$N^{\alpha\beta}_{(u)} = N^{\alpha\beta} + \tfrac{1}{2}(b^{\alpha}_{\lambda}M^{\lambda\beta} - b^{\beta}_{\lambda}M^{\lambda\alpha}),$$
$$M^{\alpha\beta} = \mathrm{M}^{(\alpha\beta)}, \qquad Q^{\alpha} = M^{\lambda\alpha}{}_{|\lambda}. \tag{4.81}$$

In these equations, the index (u) is no tensor index, but indicates that $N^{\alpha\beta}_{(u)}$ is, in general, unsymmetrical.

If we form the difference between the complete contact forces and couples $\mathrm{N}^{\alpha\beta}$, Q^{α}, $\mathrm{M}^{\alpha\beta}$ and the effective contact forces and couples (4.81), we deduce

$$\Delta M^{\alpha\beta} = \mathrm{M}^{\alpha\beta} - M^{\alpha\beta} = \mathrm{M}^{\alpha\beta} - \mathrm{M}^{(\alpha\beta)} = \mathrm{M}^{[\alpha\beta]}, \tag{4.82a}$$

$$\begin{aligned}\Delta N^{\alpha\beta} &= \mathrm{N}^{\alpha\beta} - N^{\alpha\beta}_{(u)} \\ &= \mathrm{N}^{\alpha\beta} - \mathrm{N}^{(\alpha\beta)} - \tfrac{1}{2}(b^{\alpha}_{\lambda}\mathrm{M}^{[\lambda\beta]} + b^{\beta}_{\lambda}\mathrm{M}^{[\lambda\alpha]}) - \tfrac{1}{2}(b^{\alpha}_{\lambda}\mathrm{M}^{(\lambda\beta)} - b^{\beta}_{\lambda}\mathrm{M}^{(\lambda\alpha)}) \\ &= \mathrm{N}^{[\alpha\beta]} - \tfrac{1}{2}b^{\alpha}_{\lambda}\mathrm{M}^{\lambda\beta} + \tfrac{1}{2}b^{\beta}_{\lambda}(\mathrm{M}^{(\lambda\alpha)} - \mathrm{M}^{[\lambda\alpha]}).\end{aligned} \tag{4.82b}$$

From the sixth equation of equilibrium (4.77) we find

$$e_{\alpha\beta}\mathrm{N}^{[\alpha\beta]} = e_{\alpha\beta}b^{\alpha}_{\nu}\mathrm{M}^{\nu\beta}.$$

Multiplying by $e^{\lambda\mu}$ and using $(4.13)_4$, we get

$$\mathrm{N}^{[\lambda\mu]} - \mathrm{N}^{[\mu\lambda]} = 2\mathrm{N}^{[\lambda\mu]} = b^{\lambda}_{\nu}\mathrm{M}^{\nu\mu} - b^{\mu}_{\nu}\mathrm{M}^{\nu\lambda},$$

thus

$$\mathrm{N}^{[\alpha\beta]} = \tfrac{1}{2}(b^{\alpha}_{\lambda}\mathrm{M}^{\lambda\beta} - b^{\beta}_{\lambda}\mathrm{M}^{\lambda\alpha}).$$

Substituting this expression into (4.82b) we get

$$\Delta N^{\alpha\beta} = -\tfrac{1}{2}b^{\beta}_{\lambda}(\mathrm{M}^{(\lambda\alpha)} + \mathrm{M}^{[\lambda\alpha]} - \mathrm{M}^{(\lambda\alpha)} + \mathrm{M}^{[\lambda\alpha]}) = b^{\beta}_{\lambda}\mathrm{M}^{[\alpha\lambda]}. \tag{4.82c}$$

Since the skew-symmetric tensor $\mathrm{M}^{[\alpha\beta]}$ can always be represented in the form $e^{\alpha\beta}\Phi$, where Φ is a scalar function of (θ^1, θ^2), we therefore obtain the following expressions for the *force and couple differences* (cf. (1.126)):

$$\Delta N^{\alpha\beta} = e^{\alpha\lambda}b^{\beta}_{\lambda}\Phi, \qquad \Delta M^{\alpha\beta} = e^{\alpha\beta}\Phi, \qquad \Delta Q^{\alpha} = e^{\lambda\alpha}\Phi_{,\lambda}. \tag{4.83}$$

The force and couple differences therefore depend on a single scalar function Φ. It was mentioned in section 1.5.3 that they represent internal reactions which cannot be determined uniquely within the framework of the classical theory of shells.

We now show that the force and couple differences satisfy the homogeneous equations (i.e. (4.77) with $p^{\alpha} = p^3 = 0$). Equation $(4.77)_1$ gives

$$e^{\alpha\lambda}b^{\beta}_{\lambda}\Phi_{,\alpha} + e^{\alpha\lambda}b^{\beta}_{\lambda|\alpha}\Phi - b^{\beta}_{\alpha}e^{\lambda\alpha}\Phi_{,\lambda} = 0,$$

in which Codazzi's equations (4.25) have been used. Equation $(4.77)_2$ gives

$$e^{\lambda\alpha}\Phi_{|\lambda\alpha} + e^{\alpha\lambda}b^{\beta}_{\lambda}\Phi b_{\alpha\beta} = 0,$$

since $\Phi_{|\lambda\alpha}$ as well as $b^{\beta}_{\lambda}b_{\alpha\beta}\Phi$ are symmetric in the indices α and λ. Equation $(4.77)_3$ is satisfied because this equation was used in order to define ΔQ^{α}.

Equation $(4.77)_4$ gives

$$e_{\alpha\beta}(e^{\alpha\lambda}b^{\beta}_{\lambda}\Phi - b^{\alpha}_{\lambda}e^{\lambda\beta}\Phi) = 0.$$

Since $N^{\alpha\beta}_{(u)} = \mathrm{N}^{\alpha\beta} - \Delta N^{\alpha\beta}$, etc., and the equations of equilibrium are linear in the contact forces and couples, it follows that *the effective contact forces and couples are in equilibrium with the given applied surface load* $\mathbf{p}$. We have therefore proved Theorem 1.5.2 in section 1.5.3.

4.4.4 Static–geometric analogy, stress functions

We begin by writing the homogeneous equations of equilibrium (4.75) and the compatibility conditions $(4.64)_1$ and (4.65) side by side, and we also make an unimportant change of signs in the latter equations. Thus

$$\mathbf{N}^{\alpha}{}_{|\alpha} = \mathbf{0}, \qquad (e^{\alpha\beta}\mathbf{k}_{\beta})_{|\alpha} = \mathbf{0},$$
$$\mathbf{M}^{\alpha}{}_{|\alpha} + \mathbf{a}_{\alpha} \times \mathbf{N}^{\alpha} = \mathbf{0}, \qquad (e^{\alpha\beta}\boldsymbol{\gamma}_{\beta})_{|\alpha} + \mathbf{a}_{\alpha} \times (e^{\alpha\beta}\mathbf{k}_{\beta}) = \mathbf{0}.$$

It will now be seen that the two sets of equations have the same form, so that the statical quantities $\mathbf{N}^{\alpha}$ and $\mathbf{M}^{\alpha}$ correspond to the geometrical quantities $e^{\alpha\beta}\mathbf{k}_{\beta}$ and $e^{\alpha\beta}\boldsymbol{\gamma}_{\beta}$, respectively. We also note that $\mathbf{M}^{\alpha}$ and the analogous geometrical quantities $e^{\alpha\beta}\boldsymbol{\gamma}_{\beta}$ are vectors parallel to the tangent plane, while $\mathbf{N}^{\alpha}$ and the analogous geometrical quantities $e^{\alpha\beta}\mathbf{k}_{\beta}$ may have components in the directions of all the three vectors $\mathbf{a}_{\alpha}$ and $\mathbf{a}_3$. The following analogy can therefore be established between statical and geometrical quantities (cf. section 1.5.4):

N^{α}	correspond to	$e^{\alpha\beta}\mathbf{k}_{\beta}$,
M^{α}		$e^{\alpha\beta}\boldsymbol{\gamma}_{\beta}$,
$\mathrm{N}^{\alpha\beta}$		$-e^{\alpha\lambda}e^{\beta\mu}k_{\lambda\mu}$,
Q^{α}		$e^{\alpha\beta}k_{\beta}$,
$\mathrm{M}^{\alpha\beta}$		$e^{\alpha\lambda}e^{\beta\mu}\varepsilon_{\lambda\mu}$,

in which the analogy between the components is obtained by a comparison between equations (4.73) and (4.61), i.e.

$$\mathbf{N}^{\alpha} = \mathrm{N}^{\alpha\beta}\mathbf{a}_{\beta} + \mathrm{Q}^{\alpha}\mathbf{a}_3, \qquad \mathbf{M}^{\alpha} = \mathrm{M}^{\alpha\beta}\hat{\mathbf{a}}_{\beta},$$
$$e^{\alpha\lambda}\mathbf{k}_{\lambda} = e^{\alpha\lambda}(k_{\lambda\mu}\hat{\mathbf{a}}^{\mu} + k_{\lambda}\mathbf{a}_3) = e^{\alpha\lambda}k_{\lambda\mu}e^{\mu\beta}\mathbf{a}_{\beta} + e^{\alpha\lambda}k_{\lambda}\mathbf{a}_3,$$
$$e^{\alpha\lambda}\boldsymbol{\gamma}_{\lambda} = e^{\alpha\lambda}\varepsilon_{\lambda\mu}\mathbf{a}^{\mu} = e^{\alpha\lambda}\varepsilon_{\lambda\mu}e^{\beta\mu}\hat{\mathbf{a}}_{\beta}.$$

If, therefore, we replace the statical quantities by the corresponding geometrical quantities in the homogeneous equations of equilibrium, then the resulting equations become identical with the compatibility conditions.

Corresponding to any set of *effective contact forces and couples* $N^{\alpha\beta}_{(u)}$, Q^{α}, and $M^{\alpha\beta}$ which satisfy the *homogeneous equations of equilibrium*, we shall now introduce an associated set of strain and bending measures defined by

$$N^{\alpha\beta}_{(u)} = -e^{\alpha\lambda}e^{\beta\mu}k'_{\lambda\mu},$$
$$Q^{\alpha} = e^{\alpha\beta}k'_{\beta}, \qquad M^{\alpha\beta} = e^{\alpha\lambda}e^{\beta\mu}\varepsilon'_{\lambda\mu}. \tag{4.84a}$$

(Note that the tensor $\varepsilon'_{\lambda\mu}$ defined in this manner will be symmetric because the effective contact couple $M^{\alpha\beta}$ is symmetric.) It now follows from the static–geometric analogy that these strain measures satisfy the equations of compatibility, so that a corresponding field of displacements exists. As explained in section 1.5.4, the *effective contact forces and couples* can therefore be represented in terms of *three stress functions*, which are, in fact, identical to the components of the previously mentioned displacements. Denoting the stress functions by $(\varphi_\alpha, \varphi_3)$, we replace (v_α, v_3) by $(\varphi_\alpha, \varphi_3)$ in the equations (4.56), (4.67), and (4.46), and use (4.84a) to obtain

$$\begin{aligned} N^{\alpha\beta}_{(u)} &= e^{\alpha\lambda}e^{\beta\mu}[\varphi_{3|\mu\lambda} + b^{\gamma}_{\mu|\lambda}\varphi_\gamma + b^{\gamma}_{\mu}\varphi_{\gamma|\lambda} + \tfrac{1}{2}b^{\gamma}_{\lambda}(\varphi_{\gamma|\mu} - \varphi_{\mu|\gamma})], \\ Q^{\alpha} &= e^{\alpha\lambda}e^{\beta\mu}\tfrac{1}{2}[\varphi_{\lambda|\mu\beta} + \varphi_{\mu|\lambda\beta} - 2(b_{\lambda\mu}\varphi_3)_{|\beta}], \\ M^{\alpha\beta} &= e^{\alpha\lambda}e^{\beta\mu}\tfrac{1}{2}(\varphi_{\lambda|\mu} + \varphi_{\mu|\lambda} - 2b_{\lambda\mu}\varphi_3). \end{aligned} \tag{4.84b}$$

4.5 Constitutive equations

For a thin shell consisting of a homogeneous, isotropic elastic material, the *constitutive equations for the effective contact quantities* have the form

$$\begin{aligned} N^{\alpha\beta} &= \frac{Eh}{1-\nu^2}[(1-\nu)a^{\alpha\lambda}a^{\beta\mu} + \nu a^{\alpha\beta}a^{\lambda\mu}]\varepsilon_{\lambda\mu}, \\ M^{\alpha\beta} &= \frac{Eh^3}{12(1-\nu^2)}[(1-\nu)a^{\alpha\lambda}a^{\beta\mu} + \nu a^{\alpha\beta}a^{\lambda\mu}]\kappa_{\lambda\mu}, \end{aligned} \tag{4.85}$$

and the *specific strain energy* of the shell is given by

$$W = \frac{1}{2}\frac{Eh}{(1-\nu^2)}[(1-\nu)\varepsilon^{\alpha\beta}\varepsilon_{\alpha\beta} + \nu\varepsilon^{\alpha}_{\alpha}\varepsilon^{\beta}_{\beta}] + \frac{Eh^3}{24(1-\nu^2)}[(1-\nu)\kappa^{\alpha\beta}\kappa_{\alpha\beta} + \nu\kappa^{\alpha}_{\alpha}\kappa^{\beta}_{\beta}]. \tag{4.86}$$

This may be proved in the following manner. Equations (4.85) and (4.86) are evidently tensor equations. Let us consider a point P on the middle surface, and let us choose the lines of curvature as coordinate curves. Now, the parametric representation of the surface can always be chosen in such a way that the condition $a_{\alpha\beta} = \delta_{\alpha\beta}$ is satisfied at the point P. In this special coordinate system, we have at the point P (see 1.41))

$$[a_{\alpha\beta}] = \begin{bmatrix} 1 & 0 \\ 0 & 1 \end{bmatrix}, \quad A_1 = A_2 = 1, \quad [b_{\alpha\beta}] = \begin{bmatrix} 1/R_1 & 0 \\ 0 & 1/R_2 \end{bmatrix}. \tag{4.87}$$

A comparison between the formulae (4.45) and (1.62) for $\varepsilon_{\alpha\beta}$, and the formulae (4.57) and (1.95) for $\kappa_{\alpha\beta}$ now shows that the tensor expressions for $\varepsilon_{\alpha\beta}$ and $\kappa_{\alpha\beta}$ and the expressions given in Chapter 1 for these quantities have the same values at the point P in the special coordinate system. Further, it will be seen that the formulae (4.85) and (4.86) in the special coordinate system are identical to the previous formulae (1.136) and (1.132). We have therefore shown that the tensor equations (4.85) and (4.86) are valid at the point P and in the special coordinate system. Since P is an arbitrary point on the middle surface, it follows that these equations are valid for general curvilinear coordinate systems on the surface.

4.6 Edge conditions

The contact forces and couples **n** and **m** along the boundary curve are given by the formulae (4.72) where $\boldsymbol{\nu}$ is the outward unit normal vector to the boundary curve. The contact forces and couples along the boundary curve are therefore determined by the following five quantities in terms of the complete contact forces and couples (see (4.73)):

$$\mathrm{N}^{\alpha\beta}\nu_\alpha, \qquad \mathrm{Q}^{\alpha}\nu_\alpha, \qquad \mathrm{M}^{\alpha\beta}\nu_\alpha, \tag{4.88}$$

and, as explained in section 1.7, each of these five components is equal to the corresponding component of the applied edge load.

If we use (4.82) and (4.83) to express the complete contact forces and couples as a sum of the effective contact forces and couples and the force and couple differences, we obtain the following alternative expressions for the quantities (4.88)

$$\begin{aligned} \mathrm{N}^{\alpha\beta}\nu_\alpha &= \underset{(u)}{N^{\alpha\beta}}\nu_\alpha + e^{\alpha\lambda}b^\beta_\lambda\Phi\nu_\alpha, \\ \mathrm{Q}^{\alpha}\nu_\alpha &= Q^{\alpha}\nu_\alpha + e^{\lambda\alpha}\Phi_{,\lambda}\nu_\alpha, \\ \mathrm{M}^{\alpha\beta}\nu_\alpha &= M^{\alpha\beta}\nu_\alpha + e^{\alpha\beta}\Phi\nu_\alpha. \end{aligned} \tag{4.89}$$

It was explained previously that the terms containing the function Φ represent contributions from internal reactions which cannot be determined uniquely within the framework of the theory of shells.

We shall now eliminate the function Φ from the equations (4.89). By using the formulae (see (4.30))

$$\begin{aligned} \nu^\alpha\nu_\alpha &= 1, \qquad & t^\alpha t_\alpha &= 1, \\ \nu_\alpha &= e_{\alpha\beta}t^\beta, \qquad & t^\alpha &= e^{\beta\alpha}\nu_\beta, \end{aligned} \tag{4.90}$$

we find, multiplying $(4.89)_3$ by t_β,

$$\Phi = \mathrm{M}^{\alpha\beta}\nu_\alpha t_\beta - M^{\alpha\beta}\nu_\alpha t_\beta. \tag{4.91}$$

Using $(4.72)_2$ and $(4.73)_2$, we find for the bending moment $\mathrm{M_B}$ and the torsional moment $\mathrm{M_T}$ measured per unit length of the boundary curve and originating from the complete contact forces and couples (see also Fig. 44):

$$\begin{aligned} \mathrm{M_B} &= \mathbf{m}\cdot\mathbf{t} = \mathrm{M}^{\alpha}\nu_\alpha\cdot\mathbf{t} = \mathrm{M}^{\alpha\beta}\nu_\alpha\hat{\mathbf{a}}_\beta\cdot\mathbf{t} \\ &= \mathrm{M}^{\alpha\beta}\nu_\alpha(\mathbf{a}_3\times\mathbf{a}_\beta)\cdot\mathbf{t} = \mathrm{M}^{\alpha\beta}\nu_\alpha(\mathbf{t}\times\mathbf{a}_3)\cdot\mathbf{a}_\beta \\ &= \mathrm{M}^{\alpha\beta}\nu_\alpha\boldsymbol{\nu}\cdot\mathbf{a}_\beta = \mathrm{M}^{\alpha\beta}\nu_\alpha\nu_\beta. \end{aligned} \tag{4.92a}$$

$$\begin{aligned} \mathrm{M_T} &= -\mathbf{m}\cdot\boldsymbol{\nu} = -\mathrm{M}^{\alpha\beta}\nu_\alpha\hat{\mathbf{a}}_\beta\cdot\boldsymbol{\nu} \\ &= \mathrm{M}^{\alpha\beta}\nu_\alpha(\mathbf{a}_3\times\boldsymbol{\nu})\cdot\mathbf{a}_\beta = \mathrm{M}^{\alpha\beta}\nu_\alpha t_\beta. \end{aligned} \tag{4.92b}$$

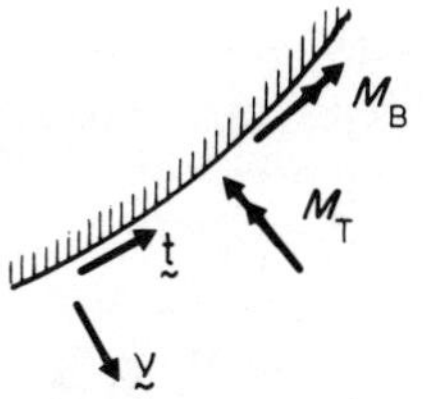

Fig. 44. Bending moment and torsional moment acting on edge

The corresponding quantities originating from the effective contact forces and couples are denoted by M_B and M_T, respectively.

From (4.91) and (4.92) we find

$$\Phi = \mathrm{M}_T - M_T. \tag{4.93}$$

Substituting this expression for Φ into the remaining equations (4.89) we find, on multiplying $(4.89)_3$ by ν_β,

$$\begin{aligned} &\mathrm{N}^{\alpha\beta}\nu_\alpha - b^\beta_\lambda t^\lambda \mathrm{M}_T = N^{\alpha\beta}_{(u)}\nu_\alpha - b^\beta_\lambda t^\lambda M_T, \\ &\mathrm{Q}^\alpha \nu_\alpha + \frac{\partial \mathrm{M}_T}{\partial s} = Q^\alpha \nu_\alpha + \frac{\partial M_T}{\partial s}, \\ &\mathrm{M}_B = M_B, \end{aligned} \tag{4.94}$$

where we have used the relation

$$\Phi_{,\lambda} t^\lambda = \frac{\partial \Phi}{\partial \theta^\lambda}\frac{\partial \theta^\lambda}{\partial s} = \frac{\partial \Phi}{\partial s}.$$

The quantities (4.94) are the four *reduced edge loads* (cf. (1.141)). We now consider a *corner point* of the boundary curve. By using (4.93) we find, since the values of the continuous function Φ before and after the corner point must be the same (cf. (1.143)),

$$\mathrm{M}_{T,\text{before}} - \mathrm{M}_{T,\text{after}} = M_{T,\text{before}} - M_{T,\text{after}}. \tag{4.95}$$

As explained in section 1.7, the statical edge conditions for the shell are formulated with the help of the reduced edge loads.

4.7 Physical components

We conclude our treatment of the theory of shells in tensor notation with an investigation of the connection between the tensor components of strain and bending measures, contact forces and couples, etc., used in the present chapter, and the components of these quantities which were used in Chapter 1.

We begin with a few remarks on physical components of tensors. Let us consider a point P on the middle surface, and let us assume that we have a system of *orthogonal curvilinear coordinates* on the surface. As explained in section 4.5, it is always possible to introduce a transformation of the independent variables with the following properties. The metric tensor $a_{\alpha\beta}$ at the point P is given by the simple expression $(4.87)_1$ in the transformed system, and the orthogonal coordinate curves are retained. It follows that the base vectors of the transformed system at the point P are mutually orthogonal unit tangent vectors $\mathbf{e}_\alpha$ to the coordinate curves. The *physical components* of a tensor at the point P are now defined as the components of the tensor in the transformed system. In the case of a vector, the physical components at the point P are therefore found by resolving the vector in the directions of the unit tangent vectors $\mathbf{e}_\alpha$ and the normal vector $\mathbf{e}_3$ (instead of using the original base vectors $\mathbf{a}_\alpha$ and the normal vector $\mathbf{a}_3$).

Denoting the physical components of tensors by *asterisks*, we find for a *vector*:

$$\mathbf{v} = \overset{*}{v}_\alpha \mathbf{e}_\alpha + \overset{*}{v}_3 \mathbf{e}_3 = v_\alpha \mathbf{a}^\alpha + v_3 \mathbf{a}_3 = v^\alpha \mathbf{a}_\alpha + v^3 \mathbf{a}_3. \tag{4.96}$$

Since

$$\begin{aligned} &\mathbf{a}_3 = \mathbf{e}_3, \qquad \mathbf{a}_\alpha = A_\alpha \mathbf{e}_\alpha, \\ &\mathbf{a}^\alpha = \frac{1}{A_\alpha} \mathbf{e}_\alpha, \end{aligned} \qquad \text{(no summation)}, \tag{4.97}$$

we obtain the following relation between the physical components of the vector and the corresponding tensor components in the original orthogonal system:

$$\begin{aligned} &\overset{*}{v}_\alpha = A_\alpha v^\alpha = \frac{1}{A_\alpha} v_\alpha, \\ &\overset{*}{v}_3 = v_3, \end{aligned} \qquad \text{(no summation)}. \tag{4.98}$$

The physical components of an *arbitrary tensor* $T^{\alpha\ldots\beta}_{\lambda\ldots\mu}$ will now be determined. We consider the invariant

$$T^{\alpha\ldots\beta}_{\lambda\ldots\mu} a_\alpha \ldots b_\beta l^\lambda \ldots m^\mu$$

at the point P, where $\mathbf{a}, \ldots, \mathbf{b}, \mathbf{l}, \ldots, \mathbf{m}$ are arbitrary vectors. We then write the expression for the invariant both in the transformed system and in the original system. Thus

$$\begin{aligned} \overset{*}{T}{}^{\alpha\ldots\beta}_{\lambda\ldots\mu} \overset{*}{a}_\alpha \ldots \overset{*}{b}_\beta \overset{*}{l}_\lambda \ldots \overset{*}{m}_\mu &= T^{\alpha\ldots\beta}_{\lambda\ldots\mu} a_\alpha \ldots b_\beta l^\lambda \ldots m^\mu \\ &= \sum_{\substack{\alpha,\ldots,\beta \\ \lambda,\ldots,\mu}} T^{\alpha\ldots\beta}_{\lambda\ldots\mu} A_\alpha \overset{*}{a}_\alpha \ldots A_\beta \overset{*}{b}_\beta \frac{1}{A_\lambda} \overset{*}{l}_\lambda \ldots \frac{1}{A_\mu} \overset{*}{m}_\mu, \end{aligned}$$

where the relation $(4.98)_1$ has been used in the derivation of the last expression. Since the vectors $\mathbf{a}, \ldots, \mathbf{b}, \mathbf{l}, \ldots, \mathbf{m}$ are arbitrary, we deduce the relation

$$\overset{*}{T}{}^{\alpha\ldots\beta}_{\lambda\ldots\mu} = \frac{A_\alpha \ldots A_\beta}{A_\lambda \ldots A_\mu} T^{\alpha\ldots\beta}_{\lambda\ldots\mu} \qquad \text{(no summation).} \qquad (4.99)$$

As P is an arbitrary point on the middle surface, this formula gives the rule of transformation connecting the physical components and the tensor components in the orthogonal curvilinear coordinate system.

We now show that the components of the displacements, strain and bending measures, contact forces and couples, etc., used in Chapter 1 are identical with the physical components of these quantities. We note that the curvilinear coordinates used in Chapter 1 constitute an orthogonal system (the lines of curvature). Introducing the notation $\bar{v}_\alpha$, $\bar{\varepsilon}_{\alpha\beta}$, $\bar{\mathrm{N}}_{\alpha\beta}$, etc., for the components used in Chapter 1, we find from (1.65), (4.43), (1.80), and (4.49a) for the displacement and rotation vectors:

$$\mathbf{v} = \bar{v}_\alpha \mathbf{e}_\alpha + \bar{v}_3 \mathbf{e}_3 = v_\alpha \mathbf{a}^\alpha + v_3 \mathbf{a}_3 = \sum_\alpha \left(v_\alpha \frac{1}{A_\alpha} \mathbf{e}_\alpha \right) + v_3 \mathbf{a}_3,$$

$$\boldsymbol{\omega} = \bar{\omega}_\alpha \hat{\mathbf{e}}_\alpha + \bar{\omega}_3 \mathbf{e}_3 = \omega_\alpha \hat{\mathbf{a}}^\alpha + \omega_3 \mathbf{a}_3 = \sum_\alpha \left(\omega_\alpha \frac{1}{A_\alpha} \hat{\mathbf{e}}_\alpha \right) + \omega_3 \mathbf{a}_3,$$

(see also (4.97)). Hence

$$\begin{aligned} \bar{v}_\alpha &= \frac{1}{A_\alpha} v_\alpha, & \bar{v}_3 &= v_3, \\ \bar{\omega}_\alpha &= \frac{1}{A_\alpha} \omega_\alpha, & \bar{\omega}_3 &= \omega_3. \end{aligned} \qquad (4.100)$$

Comparing (1.62) and (4.45), we find for the strain measures:

$$\bar{\varepsilon}_{\alpha\beta} = \frac{1}{A_\alpha A_\beta} \varepsilon_{\alpha\beta} \qquad \text{(no summation).} \qquad (4.101)$$

Comparing (1.95) and (4.57), we find for the bending measures:

$$\bar{\kappa}_{\alpha\beta} = \frac{1}{A_\alpha A_\beta} \kappa_{\alpha\beta} \qquad \text{(no summation).} \qquad (4.102)$$

It follows from sections 1.5.1 and 4.4.1 that we have the relations

$$\bar{\mathbf{N}}_\alpha = (\sqrt{a}) \mathbf{N}^\alpha, \qquad \bar{\mathbf{M}}_\alpha = (\sqrt{a}) \mathbf{M}^\alpha, \qquad (4.103)$$

or, from (1.109) and (4.73), since $\sqrt{a} = G$ for the orthogonal system of surface coordinates

$$\frac{G}{A_\alpha} \left(\sum_\beta \bar{\mathrm{N}}_{\alpha\beta} \mathbf{e}_\beta + \bar{\mathrm{Q}}_\alpha \mathbf{e}_3 \right) = G \left(\sum_\beta \mathrm{N}^{\alpha\beta} \mathbf{a}_\beta + \mathrm{Q}^\alpha \mathbf{a}_3 \right) = G \left(\sum_\beta \mathrm{N}^{\alpha\beta} A_\beta \mathbf{e}_\beta + \mathrm{Q}^\alpha \mathbf{e}_3 \right),$$

$$\frac{G}{A_\alpha} \sum_\beta \bar{\mathrm{M}}_{\alpha\beta} \hat{\mathbf{e}}_\beta = G \sum_\beta \mathrm{M}^{\alpha\beta} \hat{\mathbf{a}}_\beta = G \sum_\beta \mathrm{M}^{\alpha\beta} A_\beta \hat{\mathbf{e}}_\beta.$$

Hence,

$$\bar{Q}_\alpha = A_\alpha Q^\alpha, \qquad \text{(no summation)}.$$

$$\bar{N}_{\alpha\beta} = A_\alpha A_\beta N^{\alpha\beta}, \qquad \bar{M}_{\alpha\beta} = A_\alpha A_\beta M^{\alpha\beta}. \tag{4.104}$$

The above formulae (4.100) to (4.102) and (4.104) all agree with the rule of transformation (4.99), so that *the components used in Chapter 1 are, in fact, identical with the physical components*.

BIBLIOGRAPHY

The following is a selected list of useful works which include those referred to in the text.

Shell theory, conventional notation

1. Brøndum-Nielsen, T.: *Skalkonstruktioner* (Shell structures, in Danish). Copenhagen, 1968, 112 pp.
2. Donnell, L. H.: *Beams, plates and shells*. New York, 1976, 453 pp.
3. Flügge, W.: *Statik und Dynamik der Schalen*, 2nd edn. Berlin, 1957, 286 pp.
4. Flügge, W.: *Stresses in shells*, 2nd edn. Berlin, 1973, 525 pp.
5. Girkmann, K.: *Flächentragwerke*, 6th edn. Vienna, 1963, 632 pp.
6. Goldenveizer, A. L.: *Theory of elastic thin shells*. Oxford, 1961, 658 pp.
7. Hildebrand, F. B.: On asymptotic integration in shell theory. 'Proc. Symp. in Applied Mathematics, Vol. III. Elasticity', ed. R. V. Churchill. New York, 1950.
8. Jenkins, R. S.: *Theory and design of cylindrical shell structures*. London, 1947, 75 pp.
9. Kraus, H.: *Thin elastic shells*. New York, 1967, 476 pp.
10. Love, A. E. H.: *A treatise on the mathematical theory of elasticity*, 4th edn., Chapters 23, 24, 24A. Cambridge, 1927, 643 pp.
11. Lundgren, H.: *Cylindrical shells*, Vol. I. Copenhagen, 1949, 360 pp.
12. Novozhilov, V. V.: *Thin shell theory*, 2nd edn. Groningen, 1964, 417 pp.
13. Sanders, J. L.: An improved first approximation theory for thin shells. NASA Report 24, June 1959.
14. Seide, P.: *Small elastic deformations of thin shells*. Leyden, 1975, 654 pp.
15. Timoshenko, S., and Woinowsky-Krieger, S.: *Theory of plates and shells*, 2nd edn. New York, 1959, 580 pp.
16. Wlassow, W. S.: *Allgemeine Schalentheorie und ihre Anwendung in der Technik*. Berlin, 1958, 661 pp.

Shell theory, tensor notation

17. Bræstrup, M. W.: The Cosserat surface and shell theory. Report No. R11, Structural Research Laboratory, Technical University of Denmark, 1970, pp. 1–64.
18. Budiansky, B., and Sanders, J. L.: On the 'best' first-order linear shell theory. *Progress in Applied Mechanics, The Prager Anniversary Volume*. New York, 1963, pp. 129–140.
19. Flügge, W.: *Tensor analysis and continuum mechanics*. Berlin, 1972, 207 pp.
20. Green, A. E., and Zerna, W.: *Theoretical elasticity*, 2nd edn., Chapters 1, 10–14. Oxford, 1968, 457 pp.

21. Günther, W.: Analoge Systeme von Schalengleichungen. *Ingenieur-Archiv*, **30**, 1961, 160–186.
22. Koiter, W. T.: A consistent first approximation in the general theory of thin elastic shells. Report, Laboratory of Applied Mechanics, Technical University, Delft, 1959, pp. 1–46.
23. Koiter, W. T.: A consistent first approximation in the general theory of thin elastic shells. *The theory of thin elastic shells*, ed. W. T. Koiter. Amsterdam, 1960, pp. 12–33.
24. Koiter, W. T.: A systematic simplification of the general equations in the linear theory of thin shells. *Koninkl. Nederl. Akademie van Wetenschappen, Proceedings, Series B*, **64**, 1961, 612–619.
25. Koiter, W. T.: On the foundations of the linear theory of thin elastic shells, parts I and II. *Koninkl. Nederl. Akademie van Wetenschappen, Proceedings, Series B*, **73**, 1970, 169–195.
26. Koiter, W. T.: On the mathematical foundation of shell theory. 'Proc. Int. Congr. on Mathematics, Nice, 1970', Vol. 3. Paris, 1971, pp. 123–130.
27. Naghdi, P. M.: Foundations of elastic shell theory. *Progress in Solid Mechanics*, Vol. IV. Amsterdam, 1963, pp. 3–90.
28. Naghdi, P. M.: The theory of shells and plates. 'Handbuch der Physik', ed. S. Flügge, Vol. VI a/2, *Mechanics of Solids II*. Berlin, 1972, pp. 425–640.
29. Nielsen, M. P.: On the formulation of linear theories for thin shells. *Bygningsstatiske Meddelelser*, **35**, No. 2, 1964, pp. 37–77.
30. Niordson, F.: *Indledning til skalteorien* (Introduction to the theory of shells, in Danish). Lecture notes, Department of Solid Mechanics, Technical University of Denmark, 1974, 180 pp.

Nonlinear shell theory

31. Budiansky, B.: Notes on nonlinear shell theory. *J. Appl. Mech.*, **35**, June, 1968, 393–401.
32. Koiter, W. T.: On the nonlinear theory of thin elastic shells, parts I to III. *Koninkl. Nederl. Akademie van Wetenschappen, Proceedings, Series B*, **69**, No. 1, 1966, 1–54.
33. Mushtari, Kh. M., and Galimov, K. Z.: *Non-linear theory of thin elastic shells*. The Israel Program for Scientific Translations, Jerusalem, 1961, 374 pp.
34. Pietraszkiewicz, W.: Introduction to the nonlinear theory of shells. Report No. 10, Department of Mechanics, Ruhr Universität Bochum, May, 1977, pp. 154.
35. Sanders, J. L.: Nonlinear theories for thin shells. *Q. Appl. Math.*, **21**, No. 1, 1963, 21–36.
36. Simmonds, J. G., and Danielson, D. A.: Nonlinear shell theory with a finite rotation vector, parts I and II. *Koninkl. Nederl. Akademie van Wetenschappen, Proceedings, Series B,* **73**, 1970, 460–478.
37. Simmonds, J. G., and Danielson, D. A.: Nonlinear shell theory with finite rotation and stress-function vectors. *J. Appl. Mech.*, **39**, December, 1972, 1085–1090.

Other subjects

38. Abramowitz, M., and Stegun, I. A. (eds.): *Handbook of mathematical functions*. New York, 1965, 1046 pp.
39. Borisenko, A. I., and Tarapov, I. E.: *Vector and tensor analysis with applications*. New Jersey, 1968.
40. Copson, E. T.: *The theory of functions of a complex variable*, Chapter 10. Oxford, 1935, 448 pp.

41. Hütte. *Mathematische Formeln und Tafeln* (by I. Szabo). Berlin, 1959, 287 pp.
42. Sokolnikoff, I. S.: *Tensor analysis*, 2nd edn. New York, 1964, 361 pp.
43. Synge, J. L., and Schild, A.: *Tensor calculus*. Toronto, 1969, 324 pp.
44. Zienkiewicz, O. C.: *The finite element method*, 3rd edn. London, 1977, 787 pp.

INDEX